Things That Go Bump in the Universe

How Astronomers Decode Cosmic Chaos

C. Renée James

Johns Hopkins University Press

BALTIMORE

© 2023 Johns Hopkins University Press

All rights reserved. Published 2023

Printed in the United States of America on acid-free paper

9 8 7 6 5 4 3 2 1

Johns Hopkins University Press

2715 North Charles Street

Baltimore, Maryland 21218

www.press.jhu.edu

Cataloging-in-Publication Data is available from the Library of Congress.

A catalog record for this book is available from the British Library.

ISBN: 978-1-4214-4693-6 (hardcover)

ISBN: 978-1-4214-4694-3 (e-book)

Special discounts are available for bulk purchases of this book.
For more information, please contact Special Sales at specialsales@jh.edu.

For my parents,
who have always supported my pursuit
of shiny objects in the universe,
no matter how transient

CONTENTS

THINGS THAT GO BUMP IN THE UNIVERSE

Flashes of Insight

My fascination with fleeting, yet powerful events in the universe began during my freshman year at university. A recently matriculated physics major aiming to specialize in astrophysics, I enrolled in the usual introductory sequence expected of such a student. I entered the first astronomy class—rather haughtily—fully expecting to find out that I already knew plenty about the universe, thank you very much. I had, after all, been reading astronomy books since sixth grade, when my parents bought me the brand-new bestseller *Cosmos*, by Carl Sagan. Later, my high school physics class had introduced me to the mathematical laws that the cosmos built itself around. I felt prepared.

As it turned out, my introductory astronomy class didn't cover the material I had encountered in *Cosmos* or in any other book I had read, nor did it cover the material I had encountered in high school physics. Come to think of it, it didn't cover material I had encountered, well, anywhere. Instead, the professor professed quite a lot about the creation of elements in stellar explosions, spending only slightly less time teaching us about tiny, ghostly subatomic particles called neutrinos.

It was 1987, and my introductory astronomy professor was Donald D. Clayton. His research focus involved, as you might have guessed, the creation of elements in stellar explosions. He also had a keen interest in neutrinos, so keen that he wrote a science fiction book wherein the protagonist discovers that the production of neutrinos inside the heart of the Sun has inexplicably stopped.*

---

* Clayton's 1986 book was called *The Joshua Factor*. Spoilers: the culprit was a tiny black hole in the heart of the Sun, and it didn't turn out to be good news for Earthlings.

For all I know, Professor Clayton might have taught a typical introductory class during a typical semester. But it wasn't a typical semester, at least not astrophysically speaking. The year had featured a "nearby" stellar explosion, something that had not been witnessed since just before the invention of the telescope nearly four centuries before. "Nearby" meant that it was only about 1.6 billion billion kilometers (1 billion billion miles) away, or the distance that light can travel if you give it 168,000 years to do so. It didn't happen in our Milky Way Galaxy, but instead in the greater galactic metropolitan area in a small neighboring galaxy called the Large Magellanic Cloud (LMC). It turns out that the Large Magellanic Cloud is large only in comparison to another small neighboring galaxy falling in the same general line of sight. This second one is called, aptly, the Small Magellanic Cloud.

The stellar explosion in the LMC had been a supernova, and because it was the first of the year (it had been discovered in February 1987), it was called SN1987A. That semester, I would find out, but not fully appreciate, that SN1987A was not just a supernova by which Clayton's theoretical predictions about element creation and radioactive decays were verified, but was also the first example of interstellar multi-messenger astronomy.

You see, just hours before telescopes on Earth's surface received the light from the event, gargantuan and improbable science experiments deep underground had picked up the faintest whiff of the elusive neutrinos that heralded the stellar catastrophe. Two messengers—light and neutrinos—raced Earthward to tell us everything they knew about the death of this massive star. And if technology had progressed at a slightly different pace, the explosion might also have created detectable ripples in the fabric of the universe itself, but our ability to do that would have to wait at least three more decades.

In those intervening decades, I finished my physics degree with a hefty dose of humility, and then went on to obtain a PhD in astronomy. Rather than explore exploding stars or rippling spacetime or anything else that happens on a time scale that humans can easily imagine, I studied cosmic constants, stars that are safely entrenched in the most long-lived, most predictable part of their lives.

What I was doing was something called galactic archaeology. Like their terrestrial counterparts, galactic archaeologists try to piece together history by observing its relics. In my case, I was attempting to add something new to our understanding of the history of the Milky Way by looking at stars that preserve the chemical imprint of their birthplace. I carried out my observations using two truck-sized telescopes in West Texas, looking at the minute details of starlight with the care of an archaeologist looking at the markings on an ancient tablet.

The thing about ancient tablets, though, is that they don't explode or collide or do anything else unbecoming of an ancient tablet, and neither did the stars in my galactic archaeology project. So determined was I to make sure that the objects of my dissertation were uncontaminated by anything temporary that I excluded stars that showed even the barest hint of the presence of a companion star.

My stars needed to behave.

But the markings on ancient tablets were carved by beings that moved and lived and interacted. Sometimes those markings even recount great upheavals and terrifying cataclysms, events that changed the course of a society's history. The further I dug into the stories told by my well-behaved stars, the more I encountered an active, changing, and often violent universe. I wasn't specifically studying explosions, nor was I studying the temporary remnants of those explosions, or the subtle inward spiral of stellar dance partners, or any of a host of things that my stars' ancestors had been involved in. But I was seeing what that ancestry had wrought.

I finished my dissertation, adding a few tiles to the mosaic that illustrates how our Milky Way Galaxy got to its current state. Galactic archaeology remains a vibrant and valuable field of astronomy with more unanswered questions now than ever before. Those stories told by my stars about their ancestors, though, have kept pulling me back.

A sufficiently massive star can live for millions of years, but die in a (relative) heartbeat, its last act as bright as billions of stars put together. In death, it bequeaths to the next generation what it spent a lifetime making. Forget for a moment that this process is what helped scratch those lines on the tablets of my stars, an act for which I was academically grateful. I genuinely want to *see* a supernova! The prospect of witnessing a supernova with my own eyes is so appealing that in every class I have ever taught, I have promised that I will give an automatic A to every student if a visible supernova occurs during the semester—as though one of my students will somehow have a hotline to the universe. A long shot, to be sure, but one can hope.

Like many people, my students often think of supernovae as cosmic flashbulbs, and occasionally one will tell me with great sincerity that they saw a supernova the night before. "They're powerful and brief," I explain. "But not that brief." It takes the typical supernova months before fading into the darkness.

What *is* that brief, though, was an event that played a game of whack-a-mole with astronomers for decades. Concentrated blasts of high-energy light called gamma rays are continually popping off all over the sky, but fortunately this high-energy light is blocked by Earth's atmosphere. To spot these bursts, astronomers employ satellites designed to detect the tiny gamma ray wavelengths. Such events are so fleeting—some last less than one-tenth of a second—that for years, there were no leftover clues to help us understand their origins. I remember marveling at my colleagues in this field who wore pagers to let them know instantly when a new burst had occurred. If anyone could outsmart this sneaky opponent, they could.

Meanwhile, the subjects of my research remained steadfast and true, but I wondered . . . had *they* ever been affected by these bursts? What did the ancient inscriptions in their tablets tell us?

The late 2000s and 2010s featured some new puzzles. Instead of short-wavelength blasts, astronomers working on the other end of light's enormous spectrum were picking up brief bursts of radio waves. This news caught my eye. Radio astronomy has held a place in my scientific heart since eighth grade, when I created a science fair project devoted to the newly constructed Very Large Array, now the Karl G. Jansky Very Large Array, in Socorro, New Mexico. As its name suggests, the Karl G. Jansky VLA is indeed very large, a Y-shaped array consisting of 27 movable 25-meter dish radio telescopes. Thirteen-year-old me created a project that was, admittedly, not entirely appropriate for a science fair. It asked no research question, reported on no experiment. But it was an aesthetically pleasing and surprisingly eloquent salute to radio astronomy, complete with a small model of one of the radio telescopes crafted out of my brother's Tinkertoys, toothpicks, a starched stack of coffee filters, and a whole lot of white paint. "Is this a working model?" one judge asked hopefully. Alas . . .

When I learned of fast radio bursts—mystery blasts picked up by the much larger, but similarly shaped, Parkes Radio Telescope in Australia—my inner radio astronomer was awakened. Unfortunately, my limited knowledge of radio astronomy encompassed things that either did not change on rapid time scales or, if they did, changed predictably. Brief, regular blips from cosmic timekeepers (known as pulsars) made sense, but what could possibly spit out a single energetic radio burst? "Oh, please, let it be evaporating black holes," I implored the universe.

Not long after the discovery of fast radio bursts, gravitational wave astronomy—a field devoted almost wholly to signals that happen faster than you can say the words "gravitational wave astronomy"—took the scientific world by storm, and I enthusiastically ran to splash in its puddles. Here was something that *none* of our usual telescopes could see. Technological advances stood on the shoulders of ingenuity and creativity, and in 2015 astronomers celebrated the discovery of some of the most powerful, yet largely invisible, events in the entire universe. These colossal collisions make the very fabric of spacetime shudder.

It might seem odd that observing the equivalent of a cosmic demolition derby would require constructing the most exquisitely sensitive detectors and devising some of the most sophisticated computing algorithms, but transient astronomy is practically built on this theme.

Before I knew it, I found myself working with more and more transient astronomers.* I researched topics for a new article, a new class, and ultimately this book. Down the rabbit holes I went, dashing off this way to explore supernovae, and then that way to listen for colliding black holes, and then . . . what's that? Multi-messenger astronomy

---

* These are astronomers who study transient events; they are not temporary astronomers.

just acquired a new messenger? Amazing! The rabbit holes were being dug faster than I could run, and each month—no, each *week*—brought new discoveries and fresh insights. Not only did some of the residents in our universe change in a flash, but so did our understanding of them.

In a quiet moment of reflection, I mapped out the rabbit holes and realized that they were all part of a single warren. Moreover, it was the same one that I had unwittingly entered during my introductory astronomy class back in 1987. Now, many years later, I can say without hesitation that I do *not* know plenty about the universe, thank you very much. I have only begun to appreciate the interconnectedness of the short-lived, often violent events and the enduring, well-behaved stars—and us.

I invite you to take a leap of faith down the rabbit hole with me. You will meet the most fascinating characters and see their family tree. And with sincerest apologies to Alice, I think you'll discover that this place is more fascinating than Wonderland itself.

CHAPTER 1

Catching Cosmic Fireflies

The fireflies o'er the meadow
In pulses come and go.
—James Russell Lowell, "Midnight"

— Rock Stars —

On a cool, clear July day I waited outside an apartment building in Hornsby, a suburb a few kilometers to the north of Sydney. The wheels of trains at the nearby station tapped out a staccato rhythm, and the traffic near the local mall was beginning to gel into a glacial, slow-moving mass. Finally, a car turned into the car park, accompanied by the muffled sounds of heavy metal music. I strongly suspected that Dr. Duane Hamacher, a researcher in Aboriginal astronomy, had arrived.

Wearing a leather jacket and tweed cap, Hamacher stepped out of the car, greeted me, and helped me find a place for my backpack in the trunk. We buckled in and headed northward to Ku-Ring-Gai Chase National Park, a 37,000-acre protected natural area where, I was assured, I would encounter evidence of ancient astronomical observing.

Within minutes, we had escaped the bustle of modern life and were winding along a park road bounded by gray sandstone rocks on one side and sweeping vistas on the other. The Sun filtered through the scribbly gum trees, illuminating the crisp leaf litter on the bush-land floor. We pulled into a small car park that promised access to the Elvina Track, which would lead us part of the way to our destination. The rest of the route was a bit rough, I was told. Hamacher and I got out of the car, grabbed our packs, and began to hike up a wide, flat path.

Ku-Ring-Gai Chase National Park is situated on the large Hornsby Plateau, a great, ancient upwelling of sandstone and shale some 60 million years old. The citrus notes of eucalyptus, along with the songs of unfamiliar birds, lent an air of otherworldliness to the morning. I tried to imagine not just the astronomical observations, but also the lifestyle of the traditional Indigenous owners of Gadigal Country tens of thousands of years ago.

At a seemingly random spot on the trail, Hamacher abruptly left the marked path and began plowing through thick, crackling vegetation.

"Watch out for these," he said, indicating a spiny branch at torso level. Too late. I had already snagged my arm on one.

We weaved and bobbed our way through the scrubby brush for a few minutes, each awkward step further convincing me that we were utterly lost. Finally, we emerged onto a shallow dome of sandstone.

I saw no astronomy.

Hamacher pointed to the right of my feet, and there, carved in the rock, were two familiar shapes. "Wallabies!" I exclaimed.

He gestured to the wide sweep of stone ahead. "Careful where you step. There are dozens of carvings on this rock."

The ground was indeed covered with markings. My brain tried to separate the deliberate etchings from the odd erosion patterns. There was something fish-shaped. Carved. Several parallel lines seemed unusual, but were possibly natural. There was something elongated with stripes across it. Probably carved? A wavy outline of a person—definitely male. Two people—one male, one female—reaching up for what appeared to be a crescent Moon.

The lines were worn. In some cases, there was just a hint of a depression in the rock.

"How old are these?" I asked.

Hamacher explained that nobody knows. Five thousand years? Perhaps twice that. The English colonists had not taken great pains to communicate with the people in the area, most of whom were lost to disease and guns, both of which the English had brought to this corner of the continent.

"These are new, though," he said disdainfully, looking down. Someone, possibly even a well-meaning someone, had scratched a fresh outline on one of the figures. "People seem to think they're doing everyone a favor if they sharpen up the borders. That's why they took down the sign pointing to this site."

I now understood why it had been so important for Hamacher to get permission from Aboriginal rangers to bring me here, promising that we would do no damage.

We walked past a few more millennia-old carvings as I tried to wrap my brain around the fact that we were less than an hour's drive from a city of 5 million.

"There she is," Hamacher said, pointing.

And there she was. An enormous but slender emu in the most improbable position, her legs stretched far behind her bulbous body and her head stretched far in front. She was eight meters long from toe to beak, and as I paced the length of her figure, I saw absolutely nothing that immediately conveyed "astronomical calendar."

Although it was not obvious to me, a twenty-first-century astronomer, the Indigenous Australians knew how to read it. The emu carving has such an unusual shape not because emus of the past were wildly different, but because it mimics the morphology of the Emu in the Sky. This "constellation" is one of those things that, once seen, can never be seen as anything else. Unlike Western constellations, which create a sort of dot-to-dot map from the stars, the Emu in the Sky is sculpted out of the dark dust lanes of the Milky Way Galaxy. Her dark head is the famous Coalsack Nebula, tucked just under one arm of Crux, the Southern Cross. Her ebony neck stretches past the bright Rigel Kentaurus (formerly known as Alpha Centauri, the nearest star system to our own), and her body encompasses part of Scorpius. Aptly, deep in her belly, but some 28,000 light-years away, is the center of the Milky Way Galaxy, which hosts an enormous gravitational bottomless pit.

But none of this would have been known to astronomers in millennia past.

According to Aboriginal astronomy scholars Cilla and Ray Norris, the Emu in the Sky "stands upright above her engraving only at the time each year when the emus lay their eggs." At half a kilogram (about a pound) each, an emu egg is a veritable nutrient bomb. It is no wonder sky watchers took the time to etch this enormous shape with the correct orientation. The ancient rock carving was a link between the sky and the people. The heavens were a clock, reliable and enduring.

This reliability prompted Aristotle to pen, "In the whole range of time past, so far as our inherited records reach, no change appears to have taken place either in the whole scheme of the outermost heaven or in any of its proper parts."

That I could see the same Emu traced out in the same dark, dusty nebulae of the Milky Way Galaxy in the twenty-first century is a testament to that endurance. The aptly named Coalsack Nebula—a sooty-looking expanse—lies about 450 light-years away and contains enough material to create over 3,000 Suns. Gravity might eventually get around to pulling together stars, but the Coalsack appears strangely reluctant to do so. The Emu's head, seen by the Aboriginal people 60,000 years ago, will likely look the same 60,000 years hence. Aristotle's heavens, in other words.

Hamacher and I pushed on, eventually coming to an overlook that afforded a sweeping view of Pittwater Bay. At the far tip of the land stands the Barrenjoey Lighthouse. I imagined it steadily sweeping its beam of light, alerting ships of the dangers of venturing too close. Hamacher told me that Aboriginal people had developed their own form of lighthouse here, communicating messages and positions from far away. Naturally, as I sat on the overlook, my thoughts turned from terrestrial lighthouses to the cosmic ones known as pulsars, and I wondered aloud if the original inhabitants had witnessed an explosive stellar death that led to a pulsar.

It was a question that Hamacher and others had pondered as well. Changes in the sky had often been recorded, but not in the way that Western science records things. "There is evidence in the oral traditions of the Kokatha and Ngarrindjeri peoples that they recognized the variability of the red stars Betelgeuse, Aldebaran, and Antares," Hamacher explained. But since these records found their way into story lines rather than observation tables, their significance was overlooked until 2008, when graduate student Serena Fredrick first noted the parallels between sky stories and phenomena like Betelgeuse's variability. In 2014, Hamacher and his colleague Trevor Leaman picked up Fredrick's baton and ran farther.

Betelgeuse, a prominent star at the shoulder of Orion the hunter, does not shine steadily. It changes brightness as it swells and shrinks, brightening and dimming, over the course of about 400 days. The discovery of its variability is most commonly attributed to English astronomer Sir John Herschel, but Australia's original observers had noticed its semiregular changes long before Herschel. At its maximum brightness, Betelgeuse outshines all the other stars in Orion. At its minimum, it is reduced to ruddy mediocrity.

In 2019, Betelgeuse took a sudden and unexpected nosedive. The dimming was imperceptible from one night to the next, but over the course of a season, it lost about 60% of its usual intensity. Once solidly in the stellar top ten list, it seemed to be fading into obscurity, and by early 2020, it was noticeably dimmer even to the untrained eye.

Despite the press coverage suggesting that this fading was a potential prelude to a spectacular and explosive death—something my students and I fervently prayed for on a daily basis—the only thing that made Betelgeuse's decline impressive was the magnitude of it. It hit rock bottom in mid-February 2020 and then climbed back into its familiar rhythm once again. An obscuring blob of dust coughed out by the aging star was later determined to be the culprit. Not dead yet.

So what about other transient events in the sky? Betelgeuse might brighten and dim, but even after its 2019 free fall, it was still there. I asked Hamacher if he knew of any records, either in the rocks or in oral traditions, of one-time events. Something spectacular—say, the supernova of 1006.

There was no way Australia's Indigenous peoples could have missed it. It was the most magnificent supernova in all of recorded history, an intense pinpoint of light so bright that you could read by it. SN 1006 blazed forth in early May of that year—in the Australian autumn—just above the shoulder of the Emu. (For those familiar with Western constellations, SN 1006 appeared southwest of the constellation Scorpius in the less well-known constellation of Lupus the hare.) On the night of the new star's arrival, the crescent Moon would have quickly chased after the setting Sun, leaving the guest star to bathe this rocky overlook in a foreign, bright, blue-white light.

And not just the overlook. The entire continent of Australia and the whole *world*—at least all of it south of about 40 degrees North latitude—would have witnessed something noteworthy in the sky. For upward of three years (minus the span from October through November when it fell behind the Sun), the new star was visible, occasionally brightening and eventually dimming into the darkness from which it mysteriously sprang.

Where was its record among those whose latitude afforded them front-row seats?

— Paper Beats Rock —

The problem with understanding the information contained in rock art and story lines is twofold: (1) the information comes with virtually no context for those unfamiliar with traditional knowledge, and (2) there is quite a lot of it, and so interpreting the songs, the carvings, and the paintings can become a bit like interpreting the prophecies of Nostradamus. Scattered throughout Australia are literally thousands of sites with carvings or paintings, and much of their meaning was conveyed in symbolic narratives that have been tragically silenced. With so many centuries separating artist from observer, it is hard to divine exactly what the remaining art is telling us. Does a starburst pattern represent a supernova, a bright planetary conjunction, or a symbolic, more spiritual design with deep meaning to the artist?

For decades, I remained firmly convinced that a particular bit of Anasazi rock art in New Mexico's Chaco Canyon represented the supernova event of 4 July 1054. Carl Sagan's *Cosmos* had taught me that during my initial childhood taste of astronomy, and it likely triggered my intense yearning to see a supernova with my own eyes. "Supernovae are so amazing that people are compelled to literally carve the event in stone!" I thought.

I wanted to experience something that transformative.

But further work has cast doubt on the Chaco Canyon supernova interpretation. It might depict the event that blasted forth the Crab Nebula. Or it might not. Down the road, between Roswell and Las

Cruces, New Mexico, a similarly cryptic petroglyph depicting a wavy starburst and a scorpion *might* represent the unrivaled explosion of the supernova of 1006. Or it might not.

Only a handful of potential supernova records exists in rock art around the world, and all the astronomical interpretations are speculative at best. A Bolivian petroglyph dating back about 10,000 years could represent the explosion that resulted in a wispy cloud known as the Vela supernova remnant, while a mural on a doorway in India is thought by some to mark a stellar explosion visible in 1604, which was, by comparison, practically yesterday.

It would be arrogant to assume that cultures around the world were unaware of these punctuated changes in the heavens. Instead, what the dearth of physical evidence of cosmic fireflies seems to reveal is the overriding importance of predictability. Within a year, most visible supernovae have winked out of our skies forever. Within a century, their appearance might be woven into a greater narrative about the courage of two fishermen brothers. And within a millennium, they might not be remembered at all. Betelgeuse has the advantage of staying power, its story attached to an ongoing phenomenon.

Still, we know that there was a brilliant, temporary event in the sky that began in 1006. This knowledge is owed in large part to the almost obsessive astronomical record-keeping in the Islamic and East Asian worlds. Make no mistake, though. The custom of noting every object in the sky had less to do with scientific rigor and more to do with astrological divination. Official sky watchers were compelled to provide detailed accounts of the behavior of every object in the heavens in a vain effort to make sense of the messy business of Earthbound humans. While it is true that we owe our existence to cosmic processes, it is the rare celestial event that has an impact on our day-to-day lives.

The records of SN 1006 can be found not in the durable rocks of Australia, but in the decidedly more fragile paper of the Islamic world. "I will now describe for you a spectacle that I saw at the beginning of my education," wrote astrologer, astronomer, and physician Ali Ibn Ridwan in the eleventh century, as quoted by Bernard Goldstein. "The spectacle appeared in the zodiacal sign Scorpio in opposition to the Sun."

He recounted vividly where the spectacle appeared and how its brilliance compared to that of known objects. The account is utterly unambiguous, bested only in details by numerous reports from China. In fact, it is the Chinese chronicles that reveal the stunning duration of the supernova's visibility. SN 1006, by those records, was in the southern sky for over three years. All told, there are dozens of known accounts from China and Japan, a handful from the Arab world, and just two from Europe, where the new star would have barely peeked over the southern horizon.

And yet in all these detailed, occasionally embellished, ancient descriptions is nothing that explains what the spectacle of 1006 actually *was*.

There have been fewer than a dozen naked-eye supernovae in recorded history. Within a couple of generations of SN 1006, humanity was treated to a similar cosmic show in the constellation Taurus. That one happened on 4 July 1054, minus 6,500 or so years for light travel time, and the trigger was a single massive star that exhausted all its fuel sources and self-destructed. The explosion was so violent that the twisted, gaseous shrapnel is still racing away at over 1,000 kilometers per second. While it helps to have fancy astronomical instrumentation to measure this motion accurately, a time-lapse movie spanning a mere decade and a ruler will get you in the ballpark. In those ten years, a cosmic blink of an eye, the expansion is obvious. What is not so obvious is the little green man at its core.

There was a century-long pause between the 1054 supernova event and the next (1181), and then another four centuries before the next round of visible supernovae. The year 1572 was graced by one, as was 1604, a mere five years shy of the invention of the telescope. A supernova in our own Milky Way Galaxy has not appeared since. I tell my students that we are seriously overdue for another one, but this is a lie. Stars explode when physics dictates, not on some fixed galactic schedule. And in any event, the Milky Way *has* harbored other exploding stars, but its obscuring dust—the Emu itself—hides them from our view.

Looking deeper into history, the waters become more muddied. Chinese astronomers reported an event in the year 185 and another in

393, both of which have been declared bona fide supernovae through modern observational follow-ups. But there are a dozen more maybes, cold case files that might never see closure.

Astronomers David Green and F. Richard Stephenson suggested that it might be time to call it a day on finding ancient chronicles of exploding stars. "Looking to the future, it seems unlikely that records of additional supernovae . . . will come to light," they explained. Science historians have sifted through the sources that they can get their hands on. As for the rest? "Even to access a small proportion of this material, which is scattered in numerous archives, would be extremely time-consuming."

And what if we do run across an account of a "new star"? There is no guarantee that it describes an explosive death. A small, dense stellar corpse known as a white dwarf might suddenly and temporarily brighten if its partner encroaches upon its gravitational territory. This nonfatal flare-up is called a nova, an abbreviation of the Latin *stella nova*, which literally means "new star." Although not nearly as dramatic or as final as its cousin the supernova, a sufficiently nearby nova would make keen observers take note.

Ferreting out historical novae is a bit of a challenge, though, because they simply aren't as flashy as supernovae. A nova is a bit like a small cosmic belch. There is no epic bubble racing away from ground zero, no large-scale ionization that can be peered at centuries later with modern X-ray observatories. In a few weeks, the flare-up settles back down, and the white dwarf and its puffed-up companion go on as though nothing happened. That is, until the white dwarf reaches its limit.

Perhaps half a dozen verified naked-eye novae were recorded in the pre-telescopic era, mostly by Chinese and Japanese sky watchers, but there has been no systematic push to dig more deeply. "The observations of novae are of little value," Stephenson states bluntly in a 1976 paper about historical novae and supernovae. But "observations of supernovae . . . must surely be regarded as among the most valuable legacies which the ancient world has bequeathed to modern science."

Hamacher and I pushed back through the brushy vegetation, then across the stone, past the ancient Emu, past the wallaby sentries. If the written records of ancient astronomers are too much for modern scholars to slog through, how had he, Fredrick, and Leaman ferreted out evidence for variable stars in Aboriginal oral traditions? Fredrick's master's thesis explored the meanings of some 500 story lines, most of which have nothing to do with cosmic goings-on. It was only after spending years combing through narratives that mentioned anything in the sky and consulting with Aboriginal elders that Hamacher and Leaman were able to draw stronger conclusions about some of the less well-behaved objects like variable stars. But those are the recent stories. Tales from a millennium ago might still exist, but so far none irrefutably describe a bright new dot in the sky.

There are simply too many records to sift through. Too much data. Events that are too short-lived and too easily forgotten. These themes, as it turns out, are practically the hallmarks of studying temporary events in the universe.

"Of course," Hamacher said. "There's more to the transient universe than exploding stars and nova outbursts. Let me tell you about Collowgullouric War, which means the Wife of the Crow. But you probably know her better as Eta Carinae. Now there's a hot mess."

A hot mess, indeed. And it's destined to become an even hotter, messier mess.

But let's not go there just yet. Let's find out more about how we got here.

CHAPTER 2

Out of the Question

*We must not forget that some of the best ideas seemed
like nonsense at first.*

—Cecilia Payne-Gaposchkin, *An Autobiography*
and Other Recollections

— Hare-Raising —

On a brisk January morning in Cambridge, Massachusetts, snow flurries tickled my Texan cheeks as I made my way up Massachusetts Avenue, quickly second-guessing my choice to walk to the Harvard College Observatory from my hotel. I ducked into the bookstore, ostensibly to grab a couple of Harvard hoodies for the kids, but realistically to get out of the biting cold for a few minutes. Souvenir bag in hand, I pressed on, passing the stately redbrick buildings as I cut across a bare, crisp Harvard Yard on my way to a place I'd been aching to visit for nearly two decades.

It was the building where women had finally attained a firm foothold in astronomy, where the smudged rainbows of stars finally yielded some secrets, and where the steady heartbeats of dying stars opened our eyes to a universe much bigger and much more powerful than we had imagined.

The Harvard College Observatory was also where I would be attacked by a hare plushy.

I was ushered into the observatory building by Alison Doane, the curator of the plate stacks, who was eager to show me how it all started. Doane produced an eight-by-ten-inch pane of glass covered with dozens of smudges and handwritten notes and placed it on a light table. Handing me a jeweler's loupe, she explained that there are over half a million of these photographic plates dating back to the late

1800s. Some, like the one I was looking at, have stellar spectra on them, the light of stars spread into rainbows. But because the plates are black-and-white photographic negatives, the smudges don't look like rainbows. They are dark smears with pale stripes cutting through them, a primitive bar code of sorts. I looked through the eyepiece and found a dark streak with three prominent pale stripes. Excited to have visually classified my first A-type star, I snapped a picture of it with my iPhone and mentally prepared to insert the image into a class lecture.

Doane pulled out two more plates. Instead of spectral smudges, these were peppered with spots of varying sizes, the photographic imprints of distant stars. One plate was negative, the way it would have come out of the camera on the telescope. Bright stars appeared as dark spots on otherwise blank glass. The other plate was a positive image of the same patch of sky. Doane held one plate above the other on the light table, careful not to scrape them against each other. "Notice anything?" she asked. I didn't.

She tried to indicate a spot on the top plate, but my eyes stubbornly refused to see anything significant. "Not all the stars cancel out," she explained. "This was how Henrietta Leavitt found her variable stars."

Doane turned and picked up a cloth-bound book from the desk. "Would you like to see one of her notebooks?"

I'm pretty sure I let out a fan-girlish yelp at the question, and soon I was leafing through pages with the most immaculate columns of numbers and letters. The next thing I knew, a stuffed rabbit was bouncing on my shoulder.

"This is Henrietta Leveret," Doane laughed.

My confusion must have been written all over my face.

"A baby hare is called a leveret," she explained, somewhat disappointed. "So, this is our mascot, Henrietta Leveret."

I snapped a selfie with Henrietta Leveret and followed Doane through rows and rows of metal shelves, each shelf filled with boxes, and each box filled with photographic plates. Over half a million of them. The long and tedious process of digitizing all these plates to make the data available to the greater scientific community and the public had begun, but it would take years to electronically archive them. Walking through the plate stacks was a bit like walking through the home of a chronic hoarder whose descendants were loath to throw away anything lest it have some value down the road. No doubt, there are discoveries frozen in century-old plates that have yet to be made. The ones that have been made, though, more than make up for the clutter.

— The Data Hoarders —

In 1885, two astronomical game changers were afoot. One was the discovery of S Andromedae, a bright dot that showed up in the spiral-shaped Andromeda Nebula in August. The dot remained visible through telescopes for several months before fading out of view. Nobody knew it at the time, but it was the first telescopic supernova in history.

The other game changer was the creation of the astronomical "computers," a team of women at the Harvard College Observatory who were charged with the daunting task of making sense of more data than the observing world had ever seen.

Photography was to blame. Actually, it was photography combined with obsession. In 1885, the Harvard College Observatory had barely begun amassing the astronomical plates that I had encountered. But even as the first plates were being developed, they begged to be understood, hence the computers.

The driving force behind the "gather ye data while ye may" philosophy had been Henry Draper, a wealthy physician who, in 1872, was the first person to take a photograph of a star's spectrum. He knew that a spectrum carried in it a wealth of information, but at that point in time the scientific community didn't fully understand how to access that jackpot. Nevertheless, Draper decided that astronomers should photograph the spectra of *all the stars in the sky*, and so the Henry Draper Catalog was born.

Tragically for Draper himself, he died a decade later, but his wealth lived on at the Harvard College Observatory. Astronomers continued taking not only photographs of the rainbows produced by stars, but also repeated photographs of small patches of sky to check for any night-to-night changes in individual stars. They even sent a telescope to Peru to do the same with the stars of the Southern Hemisphere. Ultimately, the entire sky was photographed in what was the first all-sky astronomical survey. In addition, spectra of nearly a quarter of a million stars were smeared across countless panes of glass.

The photographic plates piled up.

According to Jones and Boyd, by the turn of the twentieth century, Draper's widow, Anna, started to think that perhaps this project had gone on long enough, questioning whether it was "advisable to continue to take photographs . . . night after night."

Apparently, it was.

In the 1980s, the observatory stopped adding to the storehouse of plates, not because the data taking had ceased, but because it finally became possible to record the information digitally using a combination of computers and the newfangled charge-coupled devices (CCDs). That wasn't the end of Draper's legacy, though. To this day, the Draper endowment supports the salaries of Harvard's plate stack curators, like Doane and her successors.

— Fiendishly Clever —

Enter Henrietta Leavitt. Not Leveret.

By the time Leavitt came on board as a volunteer computer in 1895, the Harvard College Observatory had already produced thousands of plates, and the corps of women computers was struggling to make sense of the dots and smudges captured on them. Leavitt had a keen eye for picking out stars that misbehaved, earning her the moniker "The Variable Star Fiend." She observed that some stars would regularly become brighter and dimmer, sometimes over a span of a day or so, and sometimes over the course of a few months.

By 1908, she had ferreted out 1,777 variable stars in two of the most prominent features in the southern sky: the Large Magellanic Cloud (LMC) and the Small Magellanic Cloud (SMC). The LMC, despite its name, is not a gaseous nebula, but a small companion galaxy to our own Milky Way. Cosmically speaking, it's practically next door. Its center lies about 163,000 light-years away from Earth, but parts of it reach much closer than that. This small galaxy stretches about 14,000 light-years across, a runt when compared to the Milky Way's

100,000-light-year diameter, but still large enough to occupy 100 square degrees of our sky. From our point of view, a fist at arm's length would just cover it. The LMC is partnered in Earth's sky with the SMC, which is both smaller and more distant than the LMC. The SMC is a mere 7,000 light-years across and lies nearly 200,000 light-years away. An outstretched fist covers the SMC many times over.

Two of our nearest satellite galaxies—they've been beaten out for the proximity award by the Sagittarius and Canis Major Dwarf Galaxies—the Magellanic Clouds are visible only to observers southward of about 20 degrees North latitude. But even at 10 degrees, you'd have to be supremely lucky to spot them. A friend of mine admits to having climbed 90 feet to the top of a tall ship's mast on the open sea to get a view of the LMC from 12 degrees North. It might sound extreme, but for a resident of the Northern Hemisphere, the chance to see entire galaxies hanging in the night sky is not one to pass up. On a moonless night away from city lights, the Magellanic Clouds pair with the sweeping arc of the Milky Way in a display of breathtaking cosmic grandeur.*

Neither of the Magellanic Clouds looks like the canonical, hurricane-shaped, swirling spiral galaxy that comes to people's minds when they hear the word "galaxy" though. The LMC, with an estimated 3 billion to 10 billion stars, is considered a "disrupted barred spiral," which means it might have been a symmetric-looking galaxy at some point, but its interaction with the much heftier Milky Way has destroyed its spiral appearance. To the unaided eye, the LMC looks like a luminous bubble of smoke spanned by a faint, glowing bar. This structure, too, will be erased in the deep future as the Milky Way gradually sips material from it through a gravitational straw. The SMC looks like a small puff of a cloud in comparison. Its official designation is "dwarf irregular galaxy," and it boasts only a few hundred million stars.

---

* I'm sure Crosby, Stills, and Nash would have made a song about the Magellanic Clouds if only the term had fewer syllables. I can hear the debate now. "No, Large Magellanic Cloud isn't going to work. Can we find a three-syllable thing? LMC? I don't think many people even know what that stands for. Let's just go with Southern Cross and be done with it."

The stars of the Magellanic Clouds come in a wide variety of sizes and ages. There are nascent stars, whose nuclear engines have only begun consuming their hydrogen fuel, and elderly stars, and every stage in between. There are regions of gas and dust compressing under gravity to form the next generation of stars, and there are the wispy remnants of the last generation. And there are the misbehaving stars.

Despite being disrupted and a bit disheveled, the Magellanic Clouds are an astronomical treasure trove. What's left of the LMC's spiral structure is face-on, giving us the fullest view of its components. Both galaxies are positioned southward of the flattened disk of the Milky Way, so our own obscuring dust doesn't mar the view. And although we don't exactly have front row seats—maybe 18th row—we do have an unobstructed view of just about any performance a star could put on. It's a show astronomers have been watching nearly non-stop for some time now, and Leavitt paid careful attention.

What Leavitt did to discover stars that changed periodically was simple and ingenious. Using the same trick that Doane showed me, Leavitt would overlay a negative plate from one night over a positive plate from another night and check for any dots that didn't cancel out.

The idea is straightforward, but the execution of it had to be mind-numbing. Imagine, hour after hour, six days a week, hunching over photographic plates with a jeweler's loupe, comparing dots and carefully documenting in endless tidy notebook columns the locations of any dots that might have changed brightness from one night to the next.* It wasn't until I held the plates and notebooks in my own hands that the sheer drudgery of the task hit me—like half a million glass panes. But what also hit me was Leavitt's superhuman determination to understand the stars.

Among the 1,777 variable stars that she spotted in the Magellanic Clouds are special stars known as Cepheid variables. For reasons

---

* These, too, have been digitized. If you want to see how our understanding of cosmic distances beyond our own backyard got its start, you can find Leavitt's notebooks here: https://library.cfa.harvard.edu/project-phaedra. If you are interested in learning more about the Harvard computers, Dava Sobel's *The Glass Universe* is outstanding.

unknown to astronomers at the time, these stars quickly—over the course of about a day—brighten by a factor of two or so, and then slowly fade back to their original brightness, only to quickly brighten and slowly fade, again and again. Leavitt found that not all Cepheids take the same amount of time to go through their cycles, nor are they all the same brightness to begin with. In a 1908 article, Leavitt commented in an oh-by-the-way fashion, "It is worthy of notice that in Table VI the brighter variables have the longer periods."

This observation was, in fact, very worthy of notice, but nobody much noticed at that point. The comment was buried deep within a paper about just a handful of Cepheid variables in the Magellanic Clouds, a subject that didn't seem to have much relevance to the Big Picture. Four years later, she cowrote a similar paper and voiced a similar sentiment, but this time she emphasized "a remarkable relation between the brightness of these variables and the lengths of their periods."

Leavitt's law, as it would later be known, was a completely unexpected relationship. All Cepheid variables that brightened and dimmed every four days, for example, seemed to have the same average energy output, while all Cepheids with a period of variability of eight days had a greater average energy output. What this means is that if you observe how long it takes for a Cepheid to brighten and dim, you can then divine what its intrinsic brightness is. This information, coupled with its apparent brightness, yields the star's distance.

The impact of Leavitt's law cannot be overstated. Before its discovery, astronomers largely relied on the triangulation method called "parallax" to get stellar distances. The principle is one people use every day. Each eye sees a slightly different view of the world, a fact that you can easily test by holding a finger close to your face and winking first one eye and then the other. Your finger appears to move back and forth as you switch eyes. For centuries, astronomers attempted unsuccessfully to see this same sort of back-and-forth motion with the stars as Earth went from one side of the Sun to the other. Seeing none of this parallax, many astronomers concluded that we weren't changing our perspective on the stars. Logically, Earth must be stationary, they argued.

The problem is that stars are so unfathomably distant that the apparent back-and-forth motion is almost nonexistent. In 1837 astronomers succeeded in measuring the minuscule parallax to a star over 11 light-years away. To appreciate how minuscule the shift was, imagine placing a cherry five kilometers (three miles) away and trying to discern its left edge from its right edge. "To determine this small shift in position," wrote astronomer Heber Curtis in 1911, "is one of the most difficult, if not absolutely the most difficult problem in astronomy, and it has taxed the utmost skill of astronomical science for three quarters of a century."

By the time Leavitt hit upon the Cepheid period-luminosity relationship, astronomers had reasonable distance estimates to only about 350 stars in the Sun's locale, and almost none of those distances were considered truly reliable. Furthermore, none of those distances were particularly large. Not cosmically speaking, anyway.

Leavitt's discovery opened the door to a much bigger, much more powerful universe.

— Transcendent Luminary —

Cepheid variables were exploited as useful tools long before anyone had any real idea what was driving them. Long before anyone had any idea what was driving *any* star, for that matter. A scientific debate raged for a few decades about whether the rise and fall of Cepheids' luminosities were tied to a pulsating mechanism, whereby the star physically got larger and brighter and then smaller and dimmer. If that were the case, what was causing such a regular pattern? And why did Leavitt's law hold true from star to star?

That debate, though, was sidelined by a much grander one. Was the Milky Way all there was to the universe? Curtis, who penned the essay about stellar distances, was convinced that the answer was no. Harlow Shapley, another prominent astronomer who had used Leavitt's law to get a rough size estimate for the Milky Way, argued otherwise. The Milky Way, by Shapley's reckoning, was simply enormous, at least tens of thousands of light-years across. Sure, the Magellanic Clouds were seemingly 100,000 light-years distant (their distance determinations were more like ballpark estimates), but they were still part of the happy Galactic family. It was folly to even consider that anything could be tens or hundreds of times farther than the Magellanic Clouds.

The unfathomable distance itself was not what gave Shapley pause, though. Parallaxes and Cepheids had already made the universe unimaginably large.

No, it was the bright new dots that showed up in the Andromeda Nebula every now and then that were most unnerving to him and to scores of other astronomers. In particular, the 1885 arrival of S Andromedae was the problem. Suddenly appearing in the hazy spiral structure and outshining the rest of the nebula put together, this dot remained visible through telescopes for several months before fading out of view. If the Andromeda *Nebula* were, in fact, the Andromeda *Galaxy*, comparable in size to the Milky Way but hundreds of thousands or even millions of light-years distant, this event "would far transcend any luminosity with which we are acquainted," Shapley wrote in 1919.

But just how much energy are we talking about? In 1890, a mere five years after the fact, Irish astronomer and science writer Agnes M. Clerke exclaimed that if S Andromedae were indeed extragalactic, "in real light, it should have been equivalent to . . . nearly *50 million such suns as our own!*"

Thirty years later, astronomers were still uncomfortable with this sort of number. Shapley concluded, "Stellar luminosities of this order seem out of the question."

Ah, yes. Definitely "out of the question." The universe just loves when astronomers say things like that.

— Stellar Surprises —

As you can see, a new light in the sky, or even just a change to a known one, has always sparked an astronomical feeding frenzy. The instant something new appears, astronomers frantically gather as much data as quickly as possible before it disappears. Of course, an astronomer spotting a new dot can't individually contact every other astronomer on the planet, certainly not in any reasonable time frame, and so we have alert systems. Or, if you prefer, BAT signals.

The first of these systems—the Central Bureau for Astronomical Telegrams (CBAT)—was created in 1882, the original clearinghouse for announcing, occasionally with actual telegrams, temporary cosmic activities. As far as anyone knew at the time, "temporary" still gave interested astronomers a few months to explore a shiny new transient. Working astronomers might get a message informing them of a "new comet" (occasionally with the less certain descriptor "new comet?"), or a "periodic comet," or even a "possible comet." There was a smattering of novae thrown in for apparent variety, "nova" being the catchall for new star-like dots that had not been previously observed. The 1885 event in Andromeda was reported via telegram as a humble nova.

The CBAT messages ultimately evolved into the International Astronomical Union Circulars, which occasionally serve as the birth announcements for bouncing baby transients. These days, the enormous breadth of astronomical research has necessitated countless specialized notification channels alerting astronomers of celestial newcomers, some of which last only a fraction of a second before winking out again.

One of the more recent alert systems is the Astronomer Telegram, a curious name given that no telegrams are actually involved. Since the dawn of the internet, the AT has allowed any bona fide member of the professional astronomical community to immediately announce a new cosmic spectacle. Unfortunately, this nearly instantaneous procedure can backfire, as cosmologist Peter Dunsby discovered in 2018. So excited was Dunsby to find "a very bright optical transient near the Trifid and Lagoon Nebulae" (AT 11448) that he immediately called on the observational community to swing their telescopes around to observe . . . Mars.

Admittedly, Mars does not stay in that particular region of the sky all the time, so it wouldn't have featured on a star chart, but the tale does underscore two important truths. First, astronomers are, as a population, genuinely kid-in-a-candy-shop excited about the possibility of witnessing something new in the sky. Second, one needs to double-check for known objects before announcing something as epic as a naked-eye transient. To his credit, Dunsby took the entire fiasco with good humor, and presumably he has turned his attention back to theoretical matters.

The first supernovae finally crept into the CBAT alerts in the 1930s, not because such things had never been observed before, but because astronomers were finally getting a better understanding of the incredible scale of the universe. In 1925, Edwin Hubble had successfully applied Leavitt's law—although he referred to it as "Shapley's period-luminosity curve"—to a dozen Cepheid variables in Andromeda and reported that the Andromeda Nebula was 930,000 light-years away. This turned out to be a pretty severe underestimate, as its currently understood distance is 2.5 million light-years, but it was enough to transform the Andromeda Nebula into the Andromeda Galaxy.

It was also enough to catapult the 1885 S Andromedae event to a new form of stardom. Once it was established that, yes, Andromeda and other "spiral nebulae" were independent galaxies, it became clear that any transient objects observed in those stupendously distant galaxies must be correspondingly stupendously bright, brighter than nearby novae by many orders of magnitude.

Now that such numbers were no longer out of the question, astronomers had to figure out how the universe could make that much energy that quickly.

CHAPTER 3

Putting the "Super" in Supernova

"Why," said the Dodo. "The best way to explain it is to do it."
—Lewis Carroll, *Alice's Adventures in Wonderland*

— Adult Supervision Recommended —

During my first teaching gig in graduate school, I nearly put out a student's eye with a pocket supernova. Honestly, the Astro Blaster didn't *seem* all that dangerous. It was just a solid rubber bouncy ball with a short plastic stalk affixed to it. Three progressively smaller bouncy balls, each with a hole drilled through it, slid over the stalk so that in the end, you had what looked like a baby's stacking toy of extremely bouncy balls. The entire apparatus was perhaps 20 centimeters (8 inches) long and weighed no more than an empty coffee mug, but the packaging exclaimed that this four-dollar wonder "works like a real supernova!"

I wasn't fully convinced that this toy would work like a real supernova, but fantasies of seeing it blast out the Earth's weight in gold ran through my head. When I got to the unit about exploding stars and the mechanisms behind them, I thought the Astro Blaster might be an amusing demonstration for my introductory astronomy students at the local community college. I took it out, threaded the three progressively smaller balls onto the stalk attached to the largest bouncy ball, and dropped it onto the classroom floor.

In retrospect, staying indoors was a mistake.

Although the whole apparatus was only dropped from the seemingly innocuous height of a meter or so, the top ball shot like a bullet into the ceiling. From there it ricocheted around the room, narrowly missing my now very-much-awake students, and coming to rest under an unoccupied desk.* I sheepishly retrieved the ball and continued the lecture, the mechanism for at least one kind of supernova now abundantly clear: a seemingly small collapse inside a star can provide enough energy to blast the outside completely off.

— Entering Neutral Territory —

Unlike the Astro Blaster manufacturer, the universe hasn't seemed particularly eager to help us on our quest to understand the power source of new dots in the sky. The telescope was invented just a few years after the last known event in our own Galaxy, and the picture in the rest of the cosmos was barely any better. By the early 1930s, only a smattering of telescopically observed transients had been recorded, and of those, precisely one (S Andromedae in 1885) had been hosted by our galactic next-door neighbor. The rest of the roster had occurred, rather inconveniently, in galaxies tens of millions of light-years away.

I say "inconveniently" because understanding something in the cosmos requires more than just spotting it. To get to the heart of these events, astronomers have to study every aspect of their light during the brief time that the objects are visible. Until very recently, distant new dots always got a sizeable head start on astronomers, who couldn't keep an eye on every galaxy every night. As a result, key information about the early days of events was inevitably lost. Sifting through the spectrum for evidence was no less challenging. Stretching out the light made it that much fainter, that much harder to collect and interpret, particularly if the object studied was 50 million light-years distant. Ideally, astronomers needed to observe where the new dot came from, what it did as the days progressed, and what, if anything, remained after the dust settled.

Still, a handful of these extragalactic events was a good enough start for astronomers Walter Baade and Fritz Zwicky. In 1934, they published the first papers about supernovae, one with the succinct and self-explanatory title "On Super-Novae." This publication coined

---

* The newest version of the Astro Blaster comes with a pair of safety goggles.

the new term, and in it, they outlined the major differences between what they termed "common novae" and "super-novae."

Common novae are, well, common. Dozens show up in our Galaxy and in other galaxies each year. Their energies are impressive, but not superlative, maxing out around 20,000 times the energy output of the Sun. Supernovae, on the other hand, are much rarer and much more energetic.

With a revised distance in hand, astronomers estimated that S Andromedae, which would become known as SN 1885A, emitted 10 million years of sunlight in under a month, far more than Clerke had imagined. As for how much total light was emitted by this event from the moment it was spotted to the moment it became invisible, Baade and Zwicky were forced to make a number of assumptions. One was that supernova progenitors are "quite ordinary stars" perhaps as much as 50 times the mass of our Sun that, for unknown reasons, suddenly expand. With a set of parameters that gave absolutely no insight into what could drive such a rapid expansion, they arrived at a value approximately equivalent to the energy produced by the Sun over *billions* of years.

More data were clearly needed. "Unfortunately, at the present time only a few underexposed spectra of super-novae are available, and it has not thus far been possible to interpret them," concluded Baade and Zwicky.

But that was not the only word the two had on the subject. The companion piece to "On Super-Novae" was the much more audacious "Cosmic Rays from Super-Novae." On the surface, the second piece concerned itself with the mysterious high-energy particles known as cosmic rays. Without much evidence—but, as it turns out, an amazing amount of prescience—Zwicky and Baade argued that these hyperactive subatomic bits originate in supernovae. Hiding in a section titled "Additional Remarks," though, was this gem: "With all reserve we advance the view that a super-nova represents the transition of an ordinary star into a *neutron star*, consisting mainly of neutrons. Such a star may possess a very small radius and an extremely high density."

A collapse mechanism had been found. The Astro Blaster could now fall.

Admittedly, the mechanism would not have been found had the neutron itself not been found. The 1932 news that English physicist James Chadwick had discovered a neutral particle in the nucleus of the atom was practically still hot off the scientific presses, but theoretically such a beast allowed great masses of material to become very small. More important, they could do so very quickly.

— Panning for Gold —

Fritz Zwicky had never intended to become an astronomer. The California Institute of Technology, better known as Caltech, had brought the Swiss physicist to Pasadena with a Rockefeller Foundation grant in 1925, and for the first few years, he largely busied himself with the study of crystal structures. With a larger-than-life personality and a nearly complete lack of a verbal filter—according to John Johnson and others, one of his favorite epithets was "spherical bastards" for those who seemed to him to be "bastards no matter which way you look at them"—Zwicky frequently challenged the old guard. Nobody was safe from his biting criticism. In 1930, he told Nobel Prize–winner and colleague Robert Millikan, who had determined the charge of an electron in a creative 1909 experiment involving tiny droplets of oil, that Millikan had never had an original idea. In response, Millikan somehow managed not to fire Zwicky. Instead, he told him to try astrophysics.

"Supernova Zwicky" was born, although the moniker "Supernova" often referred less to his success at observing supernova than to his explosive personality. He was infuriatingly open with his scorn. In a 1936 publication, he publicly mocked noted astronomer Cecilia Payne-Gaposchkin who, he stated, "failed to make the obvious and necessary distinction" between different regions in a star in her own paper as she grappled with possible spectral clues in supernovae. She had not, in fact, misunderstood.

Regardless of his personal shortcomings, Zwicky knew that he could never expect to get a real handle on supernovae with so few observations. He needed to catch more of these cosmic fireflies and study them in detail, and to do that, he needed to observe as many galaxies as possible as often as possible. By his reckoning, any given galaxy could be expected to host one supernova every millennium. Thus, by tirelessly observing 1,000 galaxies, he could anticipate spotting one supernova per year; by fixating on 10,000, he could anticipate about one per month.

The plan seems simple enough. The problem, though, is that astronomical telescopes historically have sacrificed a sweeping panoramic view for the ability to gather as much light as possible from tiny patches of sky, preferring depth over breadth. This trade-off is fine for most objects most of the time. The quiet and unchanging universe—or, more accurately, the ponderously slowly changing universe—allowed eighteenth- and nineteenth-century astronomers to catalog and map the well-behaved members of the nearby cosmos gradually and with great accuracy. Even some of the less well-behaved objects are cooperative. Henrietta Leavitt's Cepheid variables are fairly abundant, and unlike supernovae, they endure. The Cepheids that Leavitt studied are still faithfully varying in brightness, and they will continue to do so for generations to come.

What Zwicky was planning was astronomical madness, basically sacrificing endless hours in what might ultimately be a wild goose chase. What was needed to repeatedly observe thousands of galaxies was a side step in telescope technology.

The observatory at Mount Wilson, just northeast of Los Angeles, boasts a telescope that is 2.5 meters (100 inches) in diameter and the length of a truck. Although the diameter of this telescope is huge, the patch of sky it reveals is not. Different optical configurations yield different fields of view, but a monster this size can give researchers a peek at less than a millionth of the sky, or about one-hundredth of a square degree. Even if each patch of sky took only a minute, it would take years to cover it all. Worse yet, the likelihood of anything exciting and short-lived happening within that fraction of a square degree of sky during the minute it would be observed every few years is basically zero. What Zwicky needed was not a larger, more powerful telescope, but one that would capture a great chunk of the sky in one go.

The entire sky, as measured in angular units, is 41,253 square degrees. A closed fist held at arm's length takes up about 100 square degrees, about enough to cover the Large Magellanic Cloud and all its Cepheid variable stars. The view through birdwatching binoculars is surprisingly limited, revealing 20 or so square degrees, but easily encompassing the entirety of the Moon. The Moon as seen from Earth covers only about 0.2 square degrees, a figure that seems hard to believe. The Moon looks so big. And yet it would take over 200,000 of them to fill up the sky.

The human eye, despite missing out on the occasional meteor, is a marvel. It is continually surveying thousands of square degrees—an enormous swath of the available sights around you—and making note of changes. Just think of all the shiny objects that catch your eye.

Supernovae had caught Zwicky's eye, and although he had been told that searching for them was tantamount to panning for gold, he persisted. His supernova survey telescope was built atop Palomar Mountain near San Diego in 1936 and was all of 18 inches in diameter—less than half a meter—but it saw several square degrees of sky at a time. In September, he began scanning select patches of sky with high concentrations of known galaxies. Night after night, week after week, photographic plate after photographic plate, he ultimately collected nearly 10,000 "nebular images" in his search for what he would later sometimes call "atom bomb stars." During the cold winter, the irascible mountaineer and ski enthusiast built a homemade ski jump at the site. The thousands of unchanging galaxies were utterly ignored. He was interested only in finding new dots punctuating the hazy smears.

Within six months he had bagged his first supernova. Spotted on the plate from 16 February 1937, it was on the outskirts of an elongated fuzzy patch known as NGC 4157, a galaxy not too dissimilar to our own. About 100,000 light-years across, NGC 4157 lies nearly 36 million light-years from home. Since Zwicky's 1937 discovery, this galaxy has hosted two more supernova events, one observed in 1955 and one observed in 2003. As it turns out, a typical galaxy will host a supernova about twice per century, not just once per millennium.*

---

* One galaxy at 25 million light-years distant, NGC 6946, is known as the Fireworks Galaxy. This face-on spiral has featured ten supernovae in just over a century despite being considerably smaller than the Milky Way.

This was good news for Zwicky, as this higher frequency allowed him to single-handedly discover 122 supernovae, the most discovered by any one person.

Once spotted, a new supernova became the subject of intense scrutiny. At the Mount Wilson Observatory, Baade would turn the 100-inch telescope to the newcomer, insert a photographic plate, and measure the brightness of the object. Plates at Mount Wilson added up as the changing intensity of the supernova's dot was measured. The final product of all this scrutiny is known as a "light curve," a plot showing how the brightness changed over the weeks and months since the supernova's discovery.

What Baade and others soon realized is that some supernovae, after fading for a time, seem to level off in brightness for several months, only to fade away rapidly. Others quickly brighten and gradually fade away. No two supernovae are identical twins, but some definitely bear a family resemblance. Intriguingly, when astronomers accounted for the distances to the supernovae, they found that quite a large proportion of them seemed to have the same maximum brightness, give or take. Like Leavitt's law of Cepheid variable stars, this feature was flagged as a possible way to gauge colossal cosmic distances.

At the time, what light curves revealed about supernovae was not yet certain, but like detectives gathering clues, Zwicky and Baade hoped these data would add something critical to the story. They were right. Decades later, charting a light curve is still a fundamental part of observing supernovae.

By 1937, German scientist Rudolph Minkowski, escaping Nazi Germany, had joined the team. His job was to stretch out the supernova's light into a wide rainbow and comb through the various bright and dark features to find hints of the physical conditions in these events. By 1940, Minkowski realized that supernovae seem to fall into one of at least two observationally distinct categories. The spectra of some supernovae show telltale stripes at wavelengths associated with the element hydrogen, while such lines of hydrogen are absent in the spectra of others. As more spectra were obtained, astronomers began designating the supernovae with hydrogen-free spectra to be Type I. Those that showed features associated with hydrogen became Type II.

These divisions have endured through the decades, although there has been quite a bit of refinement.

What astronomers were doing then—and what many are doing now—is the celestial equivalent of the biological taxonomy studied in high school science classes. Does it have a backbone? If the answer is yes, then it's a vertebrate. Understanding the evolutionary processes that resulted in the backbone, though, is another matter entirely, one that would never have been achieved without making that first classification.

Why do some supernovae display signs of hydrogen in their spectra, while others don't? Why do some seem to level off in brightness after a while? Each new discovery seemed to spawn more questions than it answered. Still, now that supernovae were no longer once-in-a-century events, but successfully hunted quarry, the answers promised to lie just over the next hill.

CHAPTER 4

Star-Shattering Energy

Heat not a furnace for your foe so hot that it do
singe yourself.
—WILLIAM SHAKESPEARE, *HENRY VIII*

— An Earthly Perspective —

Two years before the appearance of S Andromedae, the event that
made astronomers rethink what the universe was capable of, our
planet experienced its own fabled cataclysm: the eruption of Krakatoa. The entire island was—and the operative word is "was"—only
about 45 square kilometers in area, situated inconveniently at the
boundary between the Indo-Australian tectonic plate and the Eurasian plate. As it had done for millennia, the Indo-Australian tectonic
plate crept along imperceptibly, moving scant centimeters each year
as it rode northward on a current of magma driven by forces still
deeper inside Earth's swirling interior. Off the shore of Sumatra, it hit
the Eurasian plate. Unable to resist the push of the churning interior,
but equally unable to persuade the stubborn Eurasian plate to move
out of the way, the Indo-Australian plate took a dive. Pulled beneath
the barrier by a planetary conveyor belt, it grasped at its opponent. There
was a moment of tension, just a blink of an eye on the geologic time
scale, and then . . .

Release.

It was just a bit of a slip, globally speaking. A readjustment to ease
the strain. But to the bipedal life forms scrambling around on Earth's
surface, it was legendary. One warm-up eruption on 26 August 1883
was followed the next day by four intense explosions that were quite
literally heard around the world. The force of the blasts shot ash and
pumice miles into the air, and the debris rained back down in a muddy

rock-storm. The smoke and ash darkened even the noonday skies and triggered eerie, almost supernatural lightning.

The shock waves from the explosions were so intense that anyone within 160 kilometers of ground zero was rendered permanently deaf, assuming they were lucky enough to survive. Two-thirds of the island almost immediately dropped off the map as though swallowed by a city-sized sinkhole. Tsunami after tsunami of biblical proportions— some estimate that the waves were up to 30 meters in height—swept over the nearby islands, killing tens of thousands. The weary remnants of those waves reached as far as Hawaii 11,000 kilometers (7,000 miles) to the east. Mysterious booms reminiscent of distant cannons were reported as far away as Perth, Australia, 4,500 kilometers (2,800 miles) southward. Current estimates suggest that Krakatoa's explosive power was equivalent to 200 megatons—that's 200 *million* tons—of TNT.

More subtly, barometers around the globe responded to the sudden, but imperceptible changes in air pressure as the dissipating sound waves whispered secrets about their cataclysmic origins. For days, these inaudible ripples circled the planet, nudging the readings on barometer after barometer in what was, in hindsight, a completely predictable pattern. Had meteorologists of the time been better networked, they could have even divined something of the nature of the disaster from those weak signals. But those puzzle pieces would have to wait to fall into place.

On a planet like Earth, Krakatoas are an inevitable rite of passage. The interior is a torrent of enormous convection cells that, upon reaching the surface, push and pull thin crustal fragments. Here, a rift is opened; there, a mountain range appears. And at the border of the Indo-Australian and Eurasian plates, solid crust is pulled down to a planetary forge. But fast-forward billions of years into the future, and our crust will become too thick, the convection in the mantle no match for the stubbornness of the surface.

All Krakatoas stop eventually.

Despite its fame, Krakatoa was not the most powerful event to rock our world. Not by a long shot. The dinosaur-killing asteroid impact of 65 million years ago upended the Yucatan with the strength of half a million Krakatoas, changing the course of evolutionary history in our favor. Tsunamis the height of skyscrapers would have raced across the Gulf of Mexico, chasing after winds that would make a Category 5 hurricane seem like a gentle breeze.

But as intense as these events were, they were not truly Earth-shattering.

Certainly not star-shattering.

— Fathomable Stars —

To measure what it takes to be a star shatterer, we need numbers beyond megatons of TNT, beyond Krakatoas, and even beyond the dinosaur killer. The Sun, a modest star, spends its life internalizing the equivalent of more than a billion Krakatoa eruptions *each second* in its core as it fuses hydrogen nuclei into helium nuclei, releasing energy in the process. Yet on a stellar level, this violence is easily contained, kept in check by the suffocating embrace of its 700,000-kilometer-thick (435,000-mile-thick) blanket, which pushes inward against the explosive energy with the pressure of hundreds of billions of atmospheres.

The constant distillation of energy from mass in the Sun's core, a power source that was understood only after Albert Einstein's "very interesting conclusion" in 1905 that mass and energy are equivalent, destroys nearly 4 million tons of matter every second. Taking the place of that 4 million tons per second are approximately 380 trillion trillion watts of power. Remarkably, the Sun has the capacity to eat mass at this rate and to shine with this wattage (give or take) for about 10 billion years.

The Sun's longevity can be attributed to its colossal mass. To express the Sun's heft in kilograms requires 30 zeroes, which is why astronomers almost never do that.* Instead, we simply say that it contains one solar mass. Whether reported in kilograms or solar masses, our nearby star contains more than 300,000 times as much material as Earth. The history-altering asteroid impact of 65 million years ago would have been less dramatic than a fly hitting the Sun's enormous, blazing windshield.

---

* To be more precise, the Sun contains 1.989×10^{30} kilograms (199 million trillion trillion kilograms) of material.

Thus, while consuming 4 million tons of matter per second might seem like stellar gluttony to those of us impressed by a single Krakatoa, the Sun's appetite is, cosmically speaking, more like that of a slow-nibbling picky eater. After all is said and done, the Sun will have converted just over 3% of 1% (0.034%) of its entire mass to energy. In the process, it will have released a grand total of about 120 billion quadrillion quintillion joules of energy (that's 1.2×10^{44} joules, or 12 with 43 zeroes after it). This is assuredly an enormous amount of energy, but its release is diluted over billions of years.

As you can see, attempting to describe the power source of the Sun in joules or Krakatoas becomes an exercise in keeping track of all the zeroes. But in the cosmos, these big numbers are commonplace. Baade and Zwicky first hinted at the enormity of the energies that the universe is capable of generating, and subsequent generations have refined their figures. Enormous energies are so common, in fact, that in the late 1970s, a relatively obscure unit of measure was invented almost jokingly: the foe.

As legions of astronomers had done before them, Hans Bethe and Gerald Brown were wrestling with the physics of exploding stars and their leftovers. In their calculations, Bethe and Brown frequently arrived at the staggeringly enormous 10^{51} ergs, a number similar to the original supernova calculations of Baade and Zwicky, and so they created a catchy acronym: foe = *fifty-one* *e*rgs. The unit never really caught on, unfortunately, possibly because it fails to capture the fact that 51 is *not* the number of ergs, but the power to which ten has been raised. As everyone knows, there is a vast difference between 51 mosquitoes and 10^{51} of them.*

The erg is another relatively obscure unit, and a minuscule one at that. Officially equivalent to one ten-millionth of a joule, an erg is approximately the energy required for a mosquito to take flight. If 10^{51}

---

* Because I couldn't resist, and because you are also probably trying to figure it out, you could fit only about 10^{29} *uncompressed* mosquitoes into the volume of Earth. If there were 10^{51} of them, they would fill the volume of hundreds of solar systems and pack the mass of a quadrillion Suns. They would assuredly not be uncompressed, though.

mosquitoes were to abruptly and simultaneously take flight (an impossibility even in the swampiest of climates), they would release star-shattering energy. A foe.

Why astrophysicists, who study the largest, most powerful things ever, adopted a standard unit of energy so tiny has been a perennial puzzle. It has been suggested that using the centimeter-gram-second (CGS) system of units, rather than the more familiar meter-kilogram-second (MKS, or standard metric) system simplifies some of the astrophysical equations. On the other hand, the results in either convention still become ten to the (incredibly large) power, so it really matters little whether "incredibly large" is 44 or 51.

Modern supernova researchers still agree that the quantity 10^{51} ergs (10^{44} joules) needs its own unit, though, and in those closed circles, it is known as the bethe. Truth be told, I think the foe sounds more appropriately menacing. The existence of these units illustrates what one of my graduate professors once said: The answer to every problem in astronomy is one, as long as you adopt the appropriate unit.*

The bethe (née foe) does make an excellent shorthand for swift comparisons. On that scale, Krakatoa becomes quite friendly, having exploded with only about ten trillionths of a quadrillionth of a foe, or 0.00000000000000000000000001 (10^{-26}) foes (or bethes) of energy. The Sun, meanwhile, will steadily generate a grand total of approximately 1.2 bethes of energy over its 10-billion-year lifetime, some tiny scrap of which will have been harnessed by the life forms on its third-nearest neighbor.

That's what a fathomable, well-behaved star does, after all.

— Fade to White —

Despite Zwicky's and Baade's early assertions that all stars will eventually explode, the reality—disappointing to many—is that the Sun is

---

* Astronomers seem to be continually trying to make this very point. During the writing of this book, I saw news stories about an asteroid "half the size of a giraffe" striking Earth and a type of stellar explosion that fused an amount of hydrogen equivalent to "3.5 billion Great Pyramids of Giza." There was even something reported in weasel lengths. I offer no apologies for adopting the foe.

never going to blow up. It has neither the mass nor the companion-ship to pull off this feat. While it's true that our star is fully capable of unleashing life-altering fury on us, a supernova Sun is simply not in the cards.

What is in the cards for the Sun and for countless stars containing about the same mass, give or take a factor of a few, is a lengthy run where the size and energy output remain mostly constant, at least cosmically speaking. The rest of its life will unfold fairly predictably. It will run out of fuel in its core, at which point the core, no longer expe-riencing the support of its energy source, will succumb to the inexo-rable squeeze of gravity. The core will then shrink and heat.

And shrink.

And heat.

As hydrogen pours into the reserve tank vacated by the retreating core, a new round of hydrogen fusion will kick in. This new energy, coupled with the energy released by the heating and shrinking core, will upset the balance that the Sun has enjoyed for 10 billion years. The outer layers, once equally attracted by gravity and held at bay by the churning, hot plasma inside, will feel a greater outward push than in-ward pull. The edge of the Sun will swell, and it will become a red giant, incinerating the inner planets as it does so.

It is somewhat unfair of me to characterize the Sun's life as un-eventful. Near the end, it will enjoy one brief, incredibly energetic phase, but only once the shrinking core hits the admirably hot tem-perature of 100 million degrees Celsius. At this temperature, the he-lium nuclei, each a collection of two protons and two neutrons, will find that their strong mutual repulsion is not enough to keep them from fusing into carbon nuclei, each containing six protons and six neutrons. An additional helium nucleus occasionally will find its way into one of these, and oxygen, with eight protons and eight neutrons, will be born.

Getting three of anything to come together simultaneously is challenging, but the fact that like charges strongly repel at all but the shortest of distances makes matters even worse. Overcoming this re-pulsion and thrusting the protons and neutrons together into a tightly bound carbon nucleus is a bit like trying to roll a ball up a steep hill with a deep gorge on the other side. The repulsion of the positive charges creates the hill, but there is a counterintuitive strong attraction between protons and neutrons when they get within about a proton's width of each other, hence the gorge. If you can get the particles up the hill and to the edge of the gorge, they will naturally plummet in.

The scorching hot temperatures found in the helium core of such a star are enough to race the subatomic balls up the repulsing hills—and even burrow *through* the hills. They will then collect at the bottom of the gorge as freshly made carbon nuclei.

At this point, the star experiences a bit of a crisis. Dropping a ball into a gorge releases energy, and so does combining helium nuclei into carbon. This energy finds itself trapped in the dense core, and no matter how hard it pushes on the material, it won't budge. In the everyday life of an Earthling, if something heats up, it expands. Warmer air is less dense than cooler air, a feature that allows for hot air balloons. And in the everyday life of the Sun, the same is true. But in the near-death existence of the innermost heart of the Sun, adding energy fails to expand the matter in the core. It merely heats it.

Unfortunately, heating the core means the helium nuclei are coming together even faster to form carbon as more balls are rolled up the hill and plunged into the gorge even faster. And more fusion means more heat, which means more fusion, which means more heat . . .

Which means that within a minute, the core of the Sun will experience something called a helium flash, fusing nearly half its helium to carbon and releasing as much energy in 60 seconds as it currently releases in 10 million years. Or, if you like, about 0.003 foes of energy.

It would seem that something this energetic would be enough to blow the star apart, but the process is anything but flashy, at least to an outside observer. That energy doesn't create an obvious transient event, but instead mostly goes into rearranging the core structure enough so that the helium fusion is no longer a runaway process. The dying core then settles back down, this bit of indigestion suitably soothed, and proceeds to quietly fuse the rest of its helium into carbon and oxygen. The outer portion of the star, kicked off by the initial energy imbalance, even settles back down a bit temporarily. Ultimately, though, the fragile relationship between core and envelope ends, and the outer portion of the star drifts away into space, leaving the inner core of carbon and oxygen behind.

This is no ordinary ball of carbon and oxygen, though. Gravity's unrelenting squeeze has forged a material so dense that, with its present particle residents, it can be no denser. Resisting the crush of gravity are electrons obeying the bizarre rules of the very small. On these scales, things like electrons are not really objects. An electron is happy to coexist in the exact same location as another electron, but not if the other electron is doing the exact same dance. On a subatomic, quantum level, the dances of electrons are marked by their energies, and in the dying star, gravity has compressed the matter to the point that there is no more energy for the electrons to give and still exist.

In this state, the network of carbon nuclei, oxygen nuclei, and electrons will be so dense that a teaspoon of it would weigh as much as a car. The Sun's remains will contain quite a lot of teaspoons of material: enough to fill the volume of Earth but dense enough that it will hold nearly half the mass of the Sun. Having just been immersed in a stellar forge for billions of years, the resulting object will also be white-hot. A millionth of the Sun's current volume, this dense stellar end point is known as a "white dwarf." Meanwhile, the sloughed-off outer shell will become something known as a "planetary nebula," briefly glowing—here, "briefly" means about 10,000 years—with the intense, high-energy light of the exposed heart.

No explosion. Just a dead, cooling core and a gradually exiting envelope of rarefied hot gas.

The end.

(Or is it?)

— Explosive Collapse —

Because of its mass, the Sun's fate might not be particularly exciting, but plenty of stars do explode. To create the sort of supernova that Zwicky and Baade envisioned, you need to start with a star whose mass is between 9 and 25 times that of the Sun. Those stars are not easy to come by. Less than one in 100,000 stars are born with such heft, and those that are die in a flash. If the Sun's entire 10-billion-year life were compressed into a day, a star with 10 times its mass would be gone in about three minutes. A star with 25 times its mass would be gone in less than a minute. Of the millions of stars within 1,000 light-years of Earth, there is only one monstrous 25-solar-mass cosmic mayfly—Zeta Puppis, also known as Naos—and even it is likely a hair farther than 1,000 light-years.

In the simplest explanation, the life of a star is dictated by how rapidly it uses up its own fuel stores, and this pace is determined by the unforgiving laws of physics. The nuclei of four hydrogen atoms can fuse into the nucleus of a single helium atom while converting some of the original mass to energy only in environments of extreme temperatures and pressures. The most-massive stars have such extreme environments in spades, and as a consequence they burn through their hydrogen at a rate tens of thousands times that of the Sun. If the Sun swaddles a billion Krakatoas each second in its core, these stars cradle tens of trillions. The end result is the same, though. Eventually both gluttons and dainty eaters will consume all the hydrogen on their plate (in their core), and this is where a star's mass makes all the difference.

There is a poster in nearly every Astronomy 101 classroom that illustrates the seemingly unremarkable track that the Sun and its ilk will take from hydrogen fusion to giant to planetary nebula to white dwarf. The same poster reveals the slightly more exciting fate of the one-in-a-million stars with significantly higher masses. The extreme environment that allowed for hydrogen fusion shrinks, forcing helium nuclei to join to make carbon, oxygen, neon, magnesium, sulfur, and ever heavier atomic nuclei. Each new fusion channel is shorter and shorter in duration as the star's core desperately tries to squeeze another bit of life from the nuclear mass. All the while, the dying star's outer layers are swelling, and the star morphs into a supergiant.

When the core fuses its contents into iron, the star is done. Unable to produce further energy, but equally unable to efficiently shed the energy it has created, the heart of this seething monster hits temperatures of several billion degrees Celsius and densities billions of times that of water. Although the star has spent its entire life working to create its iron core, the high-energy light trapped within now destroys it, ripping apart the iron nuclei.

It might not be immediately obvious why this should be a problem for the star, but pulling so much light energy out of the core to disintegrate the iron nuclei is like pulling out the first of many support blocks. The balance of light and particles and gravity was already a precarious one, with gravity held at bay largely by the outward push of electrons in the core. That balance is tipped ever so slightly by the removal of light energy and the rearrangement of the core's particles. The core begins to collapse, and as it collapses, it becomes hotter and denser. Soon, protons and electrons, typically holding each other at arm's length by the rules of subatomic particles, join to become neutrons. Taking all that like-charge repulsion out of the picture is like removing the last support block. The core has nothing left to hold itself up until the nuclear forces between the neutrons put a halt to the madness.

All of this plays out in less than a second. In the time between the tick and the tock, the core has compressed almost to the point of vanishing, becoming even more intensely hot and dense in the process. Now 100 billion degrees Celsius and 100 trillion times as dense as water (about 100 million times as dense as a white dwarf), this *least* extreme forge of a massive star crafts a newly minted neutron star. The energy generated in this final dramatic act of the stellar core blows the rest of the star to kingdom come in a heartbeat.

And that's what a not-so-well-behaved star can do.

— Driven —

This story makes sense at the most qualitative level. You can practically hear astronomy students muttering "iron core . . . core collapse . . . boom!" as they study for their exams. But the ellipsis between "core collapse" and "boom!" has historically been an enormous source of frustration for supernova theorists. Clearly stars *went* boom! Computer models, however, have struggled to create anything that isn't a dud.

One problem is that we simply haven't had an abundance of nearby core-collapse supernovae to scrutinize. Even by the mid-1960s, just as astronomers were getting a handle on the processes that trigger these explosions, only about 250 supernovae of *any* kind had ever been observed. Perhaps half of these had been the final acts of massive stars. Perhaps.

Things have been no easier on the theoretical side. A dying massive star is both enormous and enormously complicated. If placed where the Sun is, its outer envelope would engulf the inner planets, its interior churning with complex and dynamic interactions even before core collapse. As David Branch and J. Craig Wheeler point out in their 2017 book, *Supernova Explosions*, theorists have to contend with "hydrodynamics and turbulence, magnetic fields and rotation, weak and strong particle interactions, neutrino transport, hot and catalyzed matter at nuclear densities and above, condensed matter and many-body physics, and the potential, at least, to produce exotic 'strange' particles, never mind black holes."

It's hardly a wonder that the recipes we humans program into computers so often fail to reproduce what the universe cooks up.

The initial supernova models in the 1960s and 1970s were idealizations that by necessity ignored all but the most fundamental properties. Stars were treated as one-dimensional objects, which is to say that their internal conditions were dependent only on the distance from the center of the star. There were no asymmetries, no rotation, no magnetic fields, and certainly no hope of adequately representing reality. They are to modern supernova models what the pixelated 1972 Atari game *Pong* is to immersive virtual reality: a necessary start. You have to admit that *Pong*, for all its shortcomings, paved the way to more richly rendered environments.

For a moment, though, let's forget about all these complexities and focus on the core. The core of a massive star contains 10^{57} protons, give or take, along with an equivalent number of neutrons and electrons. Just before "the end," those core protons and electrons are forced by the extreme conditions to become neutrons. But the universal accounting system cannot let this particular union happen without an additional payment. Protons are a type of subatomic particle known as a "baryon," a relative heavyweight consisting of even tinier subatomic particles known as quarks. Electrons, on the other hand, are part of a different class known as "leptons," lightweight particles that have distinctly different interactions with the rest of the universe.

In any particle interaction, particles themselves might come and go, but some things are fixed. The universe will not allow the number of baryons to change, nor will it allow the number of leptons to change. It also strictly requires the incoming and outgoing charges to be the same. The universal ledger is extraordinarily unforgiving. What this means is that when the proton and the electron become a neutron, there must be another particle formed, another lightweight one. Furthermore, because the initial proton-electron pair had no net charge, the neutron and the newly created lepton must not have any charge. A neutral lepton *must* exist.

This is what physicist Wolfgang Pauli concluded in 1930, to his immediate regret. "I've done a terrible thing today, something which no theoretical physicist should ever do," he declared. Then he confessed the mortal sin of science: "I have suggested something that can never be verified experimentally."

Pauli understood that a neutral lepton is necessarily an elusive beast. It almost never features in high school chemistry classes alongside its well-known cousin, the electron, for the simple reason that it doesn't interact with much of anything. Ever. Chemistry is all about electrons changing their allegiances, and even the proton and neutron play secondary roles. As a result, many people are under the false impression that the material universe consists solely of protons, neutrons, and electrons. But by the time you finish reading this sentence, tens of billions of vanishingly tiny neutral leptons from deep in the heart of the Sun will have raced at nearly the speed of light through your left thumbnail. And your right knee. And your eyelids. Turn your face to the warm springtime Sun, and trillions will pass unimpeded through your visage for every beat of your heart.

Unlike the photons of sunshine warming your skin, these particles will likely never interact with any part of you (although, strictly speaking, you have about a 25% chance that *one* will affect *one* of the trillions of quadrillions of particles that comprise you at some point in your decades-long life). In fact, the entirety of Earth doesn't alter their determined straight-line trek from the Sun's core to the edge of the Galaxy and beyond, and a trillion kilometers of lead is almost as transparent to them as a perfect vacuum. Almost.

And Pauli was almost correct. The neutrino ("little neutral one" in Italian) was ultimately experimentally verified, but only because scientists devised the most enormous and bizarre traps just to catch a glimpse of a few of them.

It's hard to fathom that such a particle could play much of a role in anything noticeable. And yet, as improbable as it sounds, neutrinos are the driving force behind the explosions of massive stars. They even act as an early warning system, alerting us to the explosive brilliance that is to come, if only we can notice them in time.

— Cracking Cosmic Cold Cases —

A show as destructive as an exploding star should, it seems, leave something obvious in its wake, but unfortunately for supernova surveyors, there is no hope of seeing such a thing from millions of light-years away. The concept of an object that blew itself to smithereens did seem to account for at least one much closer enigma, though: the Crab Nebula. Astronomer Nicholas Mayall waxed eloquent in a 1939 leaflet. "In the year of our Lord 1054," he wrote with a flourish, "when Omar Khayyam was a small boy, and the Battle of Hastings was still twelve years in the future, an unknown Chinese astronomer, perhaps weary and sleepy after working all night, was astonished to see a strange and brilliant new star appear."

For nine centuries, the written record of this apparition was unknown in the West, but it surfaced in the 1920s. By then the location of the new star had been well observed, but not because it was associated with those accounts of a new star. Instead, this patch of sky had been the first entry in a catalog of objects whose common bond is their nebulous appearance. Compiled by French astronomer Charles Messier in the 1700s, the Messier Catalog is a list of 110 comet impostors, objects that have the wispy look of comets but, frustratingly to comet hunters like Messier, aren't. What he cataloged ranged from galaxies to stellar birthplaces to, in the case of Messier 1 (aka the Crab Nebula, aka NGC 1952), remnants of stellar deaths.

Technically situated in the constellation Taurus but appearing to hover over the head of Orion, the Crab Nebula is a relatively easy target for amateur astronomers with even modest telescopes. Unfortunately, through small telescopes, the object is an unremarkable fuzzy patch, which is how virtually everything in the Messier Catalog appears through such telescopes. Fuzziness was, after all, pretty much a pre-requisite for inclusion. As telescopes grew, it became apparent that there were two major categories of fuzzy patches: those that looked that way because they harbored countless stars and those that were bona fide gaseous nebulae with wispy or bubble-like appearances.

As the years passed, the Crab Nebula showed itself to be unlike anything else astronomers had seen. In short, it is a mess, a great cosmic sneeze. Irregular tendrils wind their way through a hazy, slightly elongated blob. One of the original sketches that captured anything beyond a foggy oval had the appearance of an insect, and so—inexplicably—it became known as the Crab Nebula.

What's more, astronomers noticed it was growing.

Using observations spanning decades, astronomers were able to measure this growth. By imagining the clock running in reverse, they estimated when the entire nebula was at ground zero. Edwin Hubble, best known for his work on the expansion of the entire universe, declared in 1928, "the nebula is expanding rapidly and at such a rate that it must have required about 900 years to reach its present dimensions." It seemed like far more than a coincidence that a bright new object had appeared in the same patch of sky nine centuries before.

Emboldened by this victory in the early twentieth century, astronomers scrambled to find the smoking gun remnants of other naked-eye supernovae, but those searches failed to produce anything as unambiguous as the Crab Nebula. It wasn't the scientists' fault. They just didn't have the right eyes to see them.

CHAPTER 5

The Search for Smoking Guns

> But the electromagnetic spectrum runs to zero in one
> direction and infinity in the other, so really, children,
> mathematically, all of light is invisible.
>
> —Anthony Doerr, *All the Light We Cannot See*

— In a Different Light —

It was night as I wound my way through West Virginia's Monongahela National Forest. Illuminated tree trunks and the occasional deer were the only visible parts of a countryside that was undoubtedly breathtaking during the day. I had been warned that cell service would cut out well before I reached the Green Bank Observatory, but I still felt strangely detached from the world when it did.

I pulled onto the observatory grounds, collected my keys amid several posted warnings to "PLEASE TURN OFF ELECTRONICS!" and zero admonitions to turn off my headlights, and attempted to wind down after a long day of travel. The air temperature had dropped below freezing, but that didn't stop me from wandering around in an attempt to get a good view of the stars. Maybe even a new one. Seeing even the old, familiar ones proved nearly impossible, however, as the glare from the various parking lot and building lights kept resetting my night vision. An odd design for an observatory, I thought briefly, but unlike my own eyes, the telescopes aren't bothered by visible light. They would be bothered by my cell phone, though.

I retreated to my assigned room, passing a poster that stated rather ominously "The Universe Is Whispering to Us," and went to bed, eager to see the behemoth 100-meter-wide Green Bank Telescope the next day. In the morning, I glanced out my window to find a misty autumn scene, and there in the middle of the clearing was the thing that started it all.

Well, a replica of the thing that started it all.

Before Fritz Zwicky and Walter Baade had even thought about collecting supernovae, another pioneer on the opposite coast of the United States was busily assembling what would become a telescope fully 30 meters in diameter. Karl Jansky's telescope used no lenses or mirrors, but there were a few bits and pieces of a Model T in there, along with what looked to be a horizontal partial skeleton of a construction crane. Assembled near Bell Labs in Holmdel, New Jersey, this contraption rotated as it observed, earning it the disparaging moniker "Jansky's merry-go-round." Unlike the rest of the astronomical community, he saw no need to perch his observatory high on a mountaintop. There was no point. Radio waves pass unobstructed through Earth's atmosphere.

It was 1932, and Jansky was about to become the Galileo of radio astronomy, getting a glimpse of the universe in an invisible wavelength realm that revealed more than meets the eye. There was almost no context for such observations, though. What would be the benefit of knowing that extremely long-wavelength light is apparently being emitted by the center of our Milky Way Galaxy? The visible universe is an expanding extravaganza of galaxies and nebulae and stars that occasionally explode. Photographs, spectra, and light curves had revealed this much. A visibly bright new dot on the face of a galaxy was obviously important. What wasn't obvious was the significance of the hours-long peaks and valleys of Jansky's data.

The human eye can see only a sliver of the full wavelength range of light that exists. The lengths of those visible waves are just a few hundred billionths of a meter. But light waves—or, more properly, electromagnetic waves—come in every length imaginable. For example, ultraviolet light has wavelengths just shy of those we can see. Just over the other visible edge lies infrared light, which has wavelengths longer than red. But there are shorter and longer wavelengths still. X-rays and gamma rays make up the short-wavelength end of the spectrum while microwaves and radio waves make up the long-wavelength side.

By 1930, scientists had encountered all these types of light but were not always sure how their observations tied together. Gamma rays and X-rays have mysterious, superhero names because they were first observed as surprising by-products. An 1895 experiment involving energetic beams of electrons had yielded evidence of an unknown form of energy, which became known as X-rays. Their even shorter wavelength counterparts, gamma rays, were first discovered emanating from radium in 1900, but at the time it wasn't clear that either of these was just another type of light. While it was true that X-rays could expose photographic paper, they didn't seem to do the other things that the known types of light did. In any event, the only information that X-rays and gamma rays could convey was terrestrial, since Earth's atmosphere (thankfully) shields us from these high-energy waves. There would be no X-ray or gamma ray astronomy until we could send our telescopes beyond that shield.

On the longer-wavelength side of things, the story was unfolding more deliberately. Predicted to exist in the mid-1800s, radio waves were intentionally created and detected in 1887. It's not actually as hard as it might sound. To make any kind of electromagnetic wave, all one needs to do is accelerate a charged particle. The more rapid the changes to the motion of the charged particle, the higher the frequency of the wave. Radio waves can be generated by oscillating charged particles thousands or even millions of times a second. These waves can then be detected with a device whose charged particles oscillate in response, sending their electrical signals to an apparatus that converts them to something we can sense. Sometimes, radio waves are converted to sound waves, as in the case of, well, a radio. Sometimes, they are turned into subtle electrical currents that nudge a pen poised atop a moving piece of paper, scratching out clues about the intensity of the radio emission. By the time gamma rays were being discovered, radio waves were being thoroughly exploited for long-range communications, so adept are they at racing through Earth's atmosphere.

Jansky had signed up for radio communications at New Jersey's Bell Laboratories. Even though he had published his cosmic blips in

Popular Astronomy, the astronomical implications of his strange rotating scaffold of a telescope were secondary to his quest to improve
communications. His 1933 paper, "Electrical Phenomena That Apparently Are of Extraterrestrial Origin," was, in the words of his brother,
"a wedding ceremony. It weds the science of astronomy and the science
of radio and electrical engineering."

But just as it was beginning to enjoy its honeymoon, radio astronomy came to a standstill. World War II turned all radio wave
endeavors Earthward. Radar could help spot the enemy, and radio
turned even long-distance communications into real-time exchanges.
Demand for radio engineers surged, and facilities capable of generating and receiving radio waves dotted the globe. But the young field
was still rife with technical difficulties. Unexpected noise and interference could mean the difference between a successful mission and
tragedy, and so operators tried to make sense of the interference, plan
around it, or even eliminate it. Along the way, radio experts made an
intriguing realization. Some radio interference was tied to the Sun.

And then the war ended.

— Zeroing In —

Fortunately, the radio facilities and expertise remained, and scientists
now returned their attention to the skies to determine what the Sun
had been doing to interfere with communications. In their research,
astronomers found surprisingly strong radio emissions from plenty of
objects that weren't as visibly obvious as our local star. I say "surprising" because it was often impossible to find a visual counterpart to
things that shone brightly in radio waves. The strongest radio source
outside our solar system is something in the constellation Cassiopeia,
an asterism that appears from Earth like a large W that slowly runs
tight laps around Polaris, the North Star, as Earth rotates.

Cas A, the object responsible for the radio emissions, lies 11,000
light-years distant and is so visibly faint that amateur astronomers often make it a personal challenge to spot it. With giants like the Mount
Wilson 100-inch (2.5-meter) telescope and its 200-inch (5.1-meter) big
brother, the Hale Telescope at Mount Palomar, completed in 1949,
finding Cas A should have been a cakewalk. The problem with identi-

fying a visual counterpart to the radio source, however, was knowing exactly where to look. The earliest radio telescopes lacked what is known in astronomical circles as "resolution."

To be fair, modern radio telescopes do as well. The problem lies not so much in the technology as in the laws of the universe. There is a limit to a telescope's ability to discern two different objects, and that limit depends on both the size of the telescope and the wavelength observed. Radio waves are many millions of times longer than visible light waves, so to obtain the detail seen in an optical telescope's view, a radio telescope would have to be impossibly large, Earth-sized even. Jansky had used a 30-meter-long radio antenna to sweep the sky for wavelengths that were about half the length of his antenna, and at best it revealed that one general swath of sky emitted more intense radio waves than the rest.

Radio astronomers could never hope to make a single radio telescope millions of meters long. But they could use light's wave properties to their advantage, counteracting some of the limitations that those same wave properties impose. The workaround is called "interferometry," so named because waves can interfere with each other by adding up or canceling out. By combining the outputs of two or more radio telescopes, astronomers could create the equivalent of a single radio telescope as wide as the separation between the smaller ones, at least as far as its ability to focus went. Unfortunately, this approach still only gathered as much light as the individual radio telescopes could gather, but it enabled researchers to get a view of the universe that was not simply great patches of higher and lower intensity.

While an improvement, early interferometry did not allow for pinpoint accuracy. The region around Cassiopeia looked more like a topographic map of a gentle hill than the W-shaped dot-to-dot that appears visibly to us. It was possible to tell that the radio intensity was at a maximum in a patch of sky the size of your outstretched thumb, but that was about it. It took astronomers two years to visually identify a small hazy bubble at the location of the most intense radio emission, and even longer to appreciate that bubble for what it really was. Baade and Minkowski even went so far as to declare, "There is every reason to believe that the Cassiopeia source has nothing to do with supernovae."

They had not, in fact, considered *every* reason.

By the late 1950s, astronomers had succeeded in hunting down strong radio signals at the locations of the supernova events of 1572 and 1604, bringing their total stellar murder scenes to three. But there were plenty of other similar radio signals, hinting that perhaps innumerable ghosts of long-dead stars were haunting the Milky Way Galaxy. Despite Baade and Minkowski's certainty, Cas A was soon placed in the definitely-a-supernova-remnant column, as was a ribbon-like feature in the constellation Cygnus the Swan and a ghostly bubble in the southern constellation Vela the Sails.

— Here Be Dragons —

Even with such victories, the radio sky was largely uncharted astronomical territory with few known islands. Each new correlation of a radio observation with a visible astronomical object was accompanied by a sort of "oh, yes, *now* I see it," but there were many new observations that seemed to make no sense. Astronomers tied themselves into knots as they tried to interpret what their radio telescopes were telling them.

Like any good window on the universe, radio waves urged astronomers to get a sweeping view of everything and leave the classifying and understanding for another day. Witness the birth of the First, Second, and Third Cambridge Catalogs of Radio Sources. The first, created in 1950, was a warm-up, a mere 50 entries, of which only a handful were not somehow corrupted by observational problems. By the second catalog, radio astronomers were beginning to catch their stride, and the third, published in 1959, saw solid results and astronomical names that stuck.

One of those was 3C 273.

By 1962, radio astronomers had been scratching their heads for a couple of years about the 273rd entry in the Third Cambridge Catalog of Radio Sources. It was one of a new class of things called "quasi-stellar objects": "quasi-stellar" because they looked like stars—which is to say, luminous dots—but that was where the observational similarities between stars and quasi-stellar objects ended.

The dot labeled 3C 273 had been known since about the time that the distance to the supernova in Andromeda was being debated. It was cataloged as a faint bluish star in the constellation Virgo and left at that. Overlapping it, or perhaps just nearby, was one of the brightest radio sources in the entire sky. It was hard to say if the radio source was coincident with the star-like dot because, given the nature of radio waves and telescope sizes, scientists couldn't nail down the exact location of the source. Even with the focusing tool of interferometry, the best that astronomers could determine was that the radio source was in the general area of the dot, but that wasn't good enough. They needed to know for certain.

What happened next was both serendipitous and incredibly clever. The constellation Virgo probably sounds familiar because it is one of the 12 so-called star signs, along with the constellations Taurus, Libra, Pisces, and others. Most people know Virgo because of astrology (horoscopes), not astronomy. Astronomically speaking, Virgo and the other signs of the zodiac are the ones found in the same plane in which Earth orbits the Sun. Indeed, it's the same plane that the Moon and planets can be found in, give or take. The solar system is incredibly, but not completely, flat, with all the major bodies orbiting in a vast pancake.

Far, far beyond the reaches of our solar system, but also in that same plane, are the familiar horoscope constellations. More accurately, the stars that we see in these patterns are found in the same plane. The stars of Virgo are not actually associated with each other; they just happen to lie along the same basic line of sight for Earthbound observers. The closest star is about 10 light-years away, and others are many times that distance.

If you extend the plane of Earth's orbit infinitely far into space, you will cut through the signs of the zodiac. What this means is that, from our perspective, our solar system neighbors seem to creep through the signs of the zodiac as we and they go about our orbital business. This is the reason that those constellations were deemed important. If you're looking for the Sun, Moon, or planets, look in one of those asterisms. The Sun appears to travel through the entire zodiac in a year. The Moon makes the trek in about a month.

Because the solar system is so flat—although not exactly flat—we get the opportunity to see things like eclipses as solar system bodies align. We also get to see "occultations" as something in our neighborhood passes completely in front of something more distant. The Moon, which takes up the largest patch of sky from our viewpoint, has a habit of getting in the way, frequently blocking out planets, minor planets, stars . . . basically any resident of the universe. But only if that resident lies along the plane of our solar system.

In the constellation Virgo, 3C 273 is just such a resident.

Astronomers, frustrated at their inability to definitively get a bead on the various components of this tantalizing radio source, realized that the Moon would occult 3C 273 not once, not twice, but three times in 1962. Because the motion of the Moon is so well known, observing these events would put to rest the debate over the precise locations of sources in the contour plot of this bright radio emission.

To do this, astronomers would make use of the newest and largest single radio telescope on the planet: Murriyang, which is about 350 kilometers inland from the Emu in the rock in Australia and about 1,000 years down the river of time from the day SN 1006 first appeared to Earthlings. Murriyang, a Wiradjuri name bestowed in 2020, is also known as the 64-meter Parkes Radio Telescope, an enormous single-dish receiver completed in 1961.

The Dish, as it is also known, is a spectacle. Weighing over 1,000 tons, the fully steerable paraboloid can point nearly from horizon to horizon. As it peers across the adjacent sheep paddock, the edge of its enormous bowl skims just above the ground at approximately the height of the kangaroos loping lazily around the field.*

The Dish was employed to observe the 1962 occultations in a variety of wavelengths, and the resulting plot is something that radio astronomers still regard somewhat reverently. On its own, even the impressive Dish would have been unable to pin down the position of the radio emission to the accuracy needed, but knowing exactly where the Moon's edge was and exactly what time it passed in front of 3C 273 meant that astronomers could definitively say that X marks the spot.

---

* It is also the site of the annual Parkes Elvis Festival every January.

Or rather, X, Y, and Z mark the spots. The object 3C 273 is not simply a point source, but instead has some structure to it. Regardless, once those locations were settled, it would just be a matter of using the tools of optical telescopes to get the details.

Easy.

The occultations occurred, and the Dish dished up the coordinates. The area of interest, as observed from the 200-inch Palomar telescope half a world away, contained that annoying bluish foreground star—it *had* to be a foreground star—and what appeared to be a linear, jet-like structure. Astronomers had visibly noted the "funny jet" before, so observers at Caltech were eager to get its fingerprints now that it was found to be coincident with such an intense radio source. Unfortunately, the light from the star spilled over into any spectrum they tried to obtain, and so to remedy this, they got the star's spectrum so they would know what to ignore.

What had seemed like a straightforward task became an outright conundrum. The star's spectrum was nothing like what they expected. Since before the Harvard computers had sorted stars based on the presence and strength of various features, which betrayed the makeup and physical conditions of the star, astronomers had known what to expect in a stellar bar code. Four dark stripes—indicating the presence of hydrogen—were at these four wavelengths. Sodium announced its existence at these wavelengths. Small, subtle differences in the exact positions of these wavelengths had been observed, but the differences were minuscule and easily explained by something called the "Doppler shift."

Doppler shifts let astronomers know about an object's comings and goings. If something moves toward us, all the wavelengths show up shorter than expected, much the same way that the sound of a race car has a higher pitch as it speeds toward the camera. For things moving away, the wavelengths become elongated, a phenomenon that makes the "yooooooooooo" part of a passing race car's distinct "eeeeeeeeyoooooooooooo" sound. Although not sound waves, light waves show the same properties: stretching out to longer wavelengths when the emitting object is moving away from us and foreshortening when the emitting object is moving toward us.

For stars, this stretching and foreshortening is very, very slight because the motions of stars relative to Earth are very, very small (here, "small" means only a few tens to hundreds of kilometers per second of relative speed). Sure, the 1920s had ushered in the era of large Doppler shifts, but those were reserved for distant galaxies, not nearby stars. The spectral lines of galaxies, it had been discovered, were shifted to longer wavelengths, giving them the appearance of moving away from us. Because red wavelengths are the longest visible ones, the galactic Doppler shifts to longer wavelengths are known as "redshifts." What's more, astronomers had found that the most distant galaxies have the biggest redshifts, while the closer ones have smaller redshifts. These observations, coupled with decades of theoretical cosmology, revealed an expanding universe, one where space itself seemed to be getting larger.

But not the space between stars. No, the only space that seemed to be expanding was the vast ocean of space between galaxies, which is millions and billions of light-years deep. The stars and gas of our own Milky Way Galaxy, only 100,000 light-years wide, are securely bound by gravity. Foreground stars, which are members of the Milky Way, simply did not—no, *could* not—show redshifts.

So when astronomers found that the bluish foreground star in the immediate vicinity of 3C 273 had a spectrum unlike any other star, they collectively cocked their heads in consternation. Forget the fact that it didn't even show the familiar dark stripes expected of such a star, showing bright stripes instead. The problem was that the wavelengths of the stripes were utterly unrelated to any known substance.

As the year 1962 wound to a close, astronomer Maarten Schmidt spent a good deal of time puzzling over the bizarre new star. He then decided to see if perhaps the wavelengths showed a familiar pattern but just offset, the way that the spectra of galaxies appear. Hydrogen seemed an obvious starting point, given that the simplest element makes up most of the material universe. He later wrote, "The first ratio was 1.16, the second was . . . also 1.16. It suddenly struck me that I might be seeing a redshift." Indeed, after looking at the other wavelengths, he found that the ratios were always the same. The lines that he was looking at were the same as the familiar hydrogen lines but shifted redward, each showing up at a wavelength 16% longer than it should.

It would seem that such a realization would have been a victory, but the only way to explain such a huge redshift was for the "foreground" star to be one of the most distant objects then known. Forget the piddling 2.5 million light-years to Andromeda. This thing would have to be a thousand times more distant, making the required energy output equivalent to a bit less than a foe per hour.

That's 10 billion years of solar energy over the span of a leisurely lunch.

Of course, this was clearly wrong. There was simply no way for something that bright-looking to be that distant. After all, what kind of cosmic monster could possibly power such a thing?

Unless . . .

Some astronomers suggested that a gargantuan bottomless pit—a supermassive black hole containing up to millions of times the Sun's mass—could be wreaking havoc on the heart of this distant galaxy. Devouring everything that dared to tread too close, this monster would accelerate charged particles to unfathomable speeds and temperatures. These scorching cosmic whirlpools would be more than capable of generating a supernova's energy in an hour.

But no. That was unthinkable.

It was out of the question all over again, but in a different wavelength. In fact, the objects that would later become known as "quasars" are so powerful and so concentrated that astronomers spent over a decade trying to explain them away as something else, *anything* else. Reality was simply too unbelievable, the universe just a bit too unruly, and most astronomers weren't ready for such monsters just yet.

— Whispers from the Past —

Something that radio astronomers *were* able to wrap their brains around was the brightest naked-eye supernova of all time, SN 1006. Solving that puzzle takes us back to Murriyang.

With the instrument there, astronomers Francis Gardner and Douglas Milne sought to locate the smoking gun of SN 1006 in 1965. They succeeded almost instantly, their reward an unassuming map that, were it a topographical map, would indicate a somewhat rounded valley with gentle hills on either side. To a radio astronomer, this extraterrestrial contour map displayed a now-familiar shell-like structure. By that time, bubbles associated with other supernova remnants were relatively well known, and this one fit right in. The object was christened PKS 1459-51 ("PKS" for Parkes and "1459-51" to represent its celestial coordinates), and a call to find its optical counterpart was made. It would be two decades before astronomers would find even a hint of visible light in the remains of the brightest cosmic event ever recorded.

Some parts of the universe were indeed whispering to us, but despite their quietness, they made far more sense than the shrieks of 3C 273.

— The Heart of the Matter —

There remained the uncomfortable question of neutron stars.

They were allegedly the other leftovers of the unfathomable explosions that gave rise to the Crab Nebula and other wispy objects, so where were they all hiding? By all reckonings, the Milky Way should be practically swimming in these ridiculously dense objects. Millions of them. Tens of millions. Maybe even hundreds of millions. With the size of a city, but the mass of up to a million Earths, a neutron star begins where a massive star ends, and plenty of massive stars have ended over the nearly 14-billion-year history of the universe.

And yet three decades after Baade and Zwicky had hypothesized them as the required trigger behind supernovae, neutron stars—with the possible exception of the tiny off-center dot in the Crab Nebula— remained stubbornly hidden.

No matter. Astronomers could still refine their theoretical descriptions of the objects. It was agreed that neutron stars must have stupendously high densities, easily 100 trillion times that of water, so massive and so compact that these city-sized balls are able to spin hundreds of times per second and not fly apart. They also possess magnetic fields that scientists in Earthly labs could only dream of creating. More than a trillion times stronger than that of Earth, the magnetic fields of neutron stars are able to outdo our most extreme magnetic achievements by more than a factor of a million. A nascent neutron star also has a fever of a trillion degrees Celsius or so, but once free of the crushing weight of the rest of the massive star, it cools remarkably rapidly. Within a few seconds, its temperature drops to only a few hundred billion degrees, and within a few years its surface cools to a more temperate million or so degrees. But then the cooling proceeds more gradually, and even after billions of years—the age of the observable universe—a neutron star is still hotter than the Sun.

Neutron stars represent some of the most extreme environments that the universe can conjure up. And yet they did not show us the most extreme sights. Instead, they announced their presence with a "bit of scruff."

CHAPTER 6

Detecting Cosmic Heartbeats

The only true voyage of discovery . . . would be not to visit strange
lands but to possess other eyes, to behold the
universe through the eyes of another.
—Marcel Proust, *The Captive*

— Not Saying It Was Aliens . . . —

In July 1967, Jocelyn Bell (later Bell Burnell) was again scouring the printouts of her new radio telescope's data and noticed a quarter-inch-wide signal that barely rose above the background noise. "A bit of scruff," she called it. While the signal might have been all too easy for anyone else to ignore, she was determined. She had practically single-handedly strung 120 miles of wire to create the 4.5-acre telescope at Cambridge University's Mullard Radio Astronomy Observatory, a telescope that looked more like the skeleton of a vineyard than an astronomical tool. This effort made her fast friends with everything this oddly shaped telescope ever picked up. She was not about to let anything escape her attention, no matter how scruffy.

Fortunately, the scruff showed up more than once. As she pored over the miles of charts, the keen-eyed graduate student spotted the signal on about 10% of the printouts, four minutes earlier each day. Since it was keeping time with the stars, whose rise times are thrown slightly off from the Sun's by a combination of Earth's rotation and revolution, the source was definitely not terrestrial. But what was it? She persuaded her PhD advisor, Antony Hewish, to let her speed up the paper feed so the odd signals could be better scrutinized. For weeks, miles of charts streamed through, and as the piles of paper grew, so did Hewish's exasperation. Finally, on 28 November 1967, as they were about to pull the plug on the search, the signal returned.

Now, instead of a bit of fuzz, a series of regularly spaced bumps appeared, each lasting just a few tens of milliseconds and separated from its neighbors by 1.3373 seconds. These precisely timed, rapid radio blips telegraphed information about uncharted astrophysical territory. Because known stars are incapable of changing brightness so rapidly, the researchers knew they were looking at an unfathomably dense, small object, or perhaps—did they dare suggest?—it was something artificial.

Most likely not artificial, they agreed. One can imagine the uneasy laughter that accompanied the discussion between graduate student and supervisor as this thought came out in the open. Still, unusual signals are a siren song to astronomers, and despite a healthy skepticism, most are open to the possibility that something in this vast universe might try to make contact. Bell Burnell gave the object the tongue-in-cheek nickname of LGM-1 (LGM = "little green men").

If there were indeed little green men, she reasoned, they would live on a planet orbiting a star. That planet would spend part of its orbit heading away from Earth and another part of its orbit heading toward us. To and fro, to and fro, as the years went by. The next logical step was to look for Doppler shifts, those telltale changes in wavelength that let astronomers know about an object's comings and goings. These new data would inform them if the signal had been broadcast from a planet orbiting its star.

Before that investigation even got off the ground, though, LGM-1 was joined by another similarly regular signal.

Then another.

Then another.

At that point, Jocelyn Bell Burnell recalled, they gave up on aliens. "It was extremely unlikely that there would be four separate lots of Little Green Men," she explained in an email to me, "all, at the same time, signaling to the inconspicuous planet Earth, using a stupid frequency and a daft technique."

— Stranger than Aliens —

It might not have been an extraterrestrial greeting, but it was a tremendous astronomical discovery in its own right. The astronomers soon found out that whatever it was kept time to better than one part in 10 million, either rotating or pulsating—at the time, they weren't sure which—over 60,000 times per day. On top of this, it was emitting enough power across interstellar distances to be detected by the 1967 radio telescopes. This sort of activity requires a phenomenal source of energy, the one Zwicky and Baade had hypothesized over 30 years earlier. That previous generation of astronomers had resigned themselves to the fact that neutron stars were a wild idea, conveniently solving the problem of a supernova's trigger but inconveniently offering little chance of observational verification. After all, how would a city-sized ball of neutrons, even a million-degree one, ever be detected?

Then came LGM-1, which soon became CP1919 and ultimately PSR B1919 + 21. If the object were physically pulsing, growing and shrinking, Bell Burnell and Hewish computed that it would have a density of 10 trillion grams per cubic centimeter, a mountain squeezed into a thimble, or essentially the same density as a ball of neutrons. If, on the other hand, it were rotating once every 1.3373 seconds, it would need to be similarly dense to keep from flying apart. Either way, since it was a *puls*ating signal from a stell*ar* source, the object was christened a "pulsar" in 1968 by prolific science journalist Anthony Michaelis. Although scientists were still several months from understanding whether the object was pulsating or rotating, the moniker stuck.

It was admittedly a catchy name. Less than three years after Bell Burnell's discovery, the Hamilton Watch Company began developing the Pulsar watch, capitalizing on nature's seemingly perfect timekeepers. The year 1978 saw the rollout of the first Nissan Pulsar, which was—what else?—a *compact* car. In 1979, pulsars even slipped effortlessly and largely unrecognized into popular culture. Next to the entry for "pulsars" in the 1977 edition of the *Cambridge Encyclopedia of Astronomy* had been an image of successive stacked pulse profiles from PSR B1919 + 21. The simple but enigmatic portrait of squiggles might have remained in that encyclopedia and in obscure journal articles, but the image caught the eye of Peter Saville, who decided that it would make the perfect album art for Joy Division's *Unknown Pleasures*. These days, plenty of people recognize the image, which has been featured on everything from T-shirts to tattoos, but few realize it represents the first pulsar ever discovered.

Or was it?

— The Accidental Astronomers —

In July 1967, US Air Force staff sergeant Charles Schisler realized that an odd signal had been showing up week after week. The signal kept time with the stars, rising four minutes earlier each night. Anyone else might have ignored it, but Schisler was an amateur astronomer, and he understood what the timing meant.

Not that it mattered. A signal from the stars wasn't anything his employer was interested in. You see, the Ballistic Missile Early Warning System at Clear Air Force Station in Alaska was built to detect incoming warheads, not pulsars. As a result, he wasn't at liberty to announce his discovery or do any formal follow-up observations. Informally he did follow up, though, and in 2007, when information from the early warning system finally became declassified, Schisler revealed that he had observed not just one pulsar, but a dozen or so.

None of this should be a surprise. The military hadn't abandoned their use of radio waves once World War II ended, and they had a budget for technology that far outstripped what Antony Hewish and Jocelyn Bell Burnell dreamed of. The military also would have happily adopted any and all novel astronomical technology that they deemed useful. Anything to keep closer tabs on the enemy.

Radio waves were only part of the story, though. By 1967, with the global proliferation of nuclear weapons, governments were monitoring each other's activities not just with ground-based stations, but also with spacecraft. With nearly a decade of successful satellite launches, along with a decent run of manned missions, it was practically inevitable that both the United States and the USSR would want to position new eyes in space.

These were not radio eyes, though, catching a glimpse of very low-energy transmissions. These were eyes that could look for gamma rays and X-rays, the most energetic types of light in the universe. In the 1960s, as far as anyone knew, the only way to create a flood of such high-energy radiation was to detonate a nuclear bomb. While certain testing was allowed, there had been an international agreement that testing nuclear weapons in the atmosphere or in space was strictly forbidden. Of course, this rule could be broken, and the only way to know for certain was to position some satellites in orbit and be on the lookout for intense bursts of high-energy light.

And that's what the *Vela* satellites did.

Looking a bit like a cross between a disco ball and a 20-sided *Dungeons and Dragons* die, each of these 2-meter-wide satellites could detect, but not precisely localize, gamma rays. Making a gamma ray telescope is not as straightforward as creating a visible light or radio telescope. Unlike the lower-energy waves, gamma rays don't bounce off a curved mirror and focus at a point. They just plow into the detector material, losing much of their energy in the process. The best any given sensor can say is that it picked up a gamma ray, but it can't tell you where it came from.

There are a few workarounds to this inconvenient fact. One solution is to create something called a "collimator," which is a tube-like device that allows only gamma rays from a very specific direction to make it to the detector while the rest are absorbed by the tube's walls. But such a configuration wasn't really necessary for the *Vela* satellites. Scientists knew what the gamma ray signal of a nuclear bomb was, so it was just a matter of looking for one. Ultimately the *Vela* spacecraft wound up being an array of detectors. With some precise timing, it was possible to determine the order in which the detectors were triggered and then figure out, at least generally, the direction of the gamma ray source.

And so, between 1963 and 1965, the United States had launched a dozen *Vela* satellites to keep a watchful eye on our neighbors over the coming years, never suspecting that they were keeping an eye on the universe as well. The first surprise came in 1967, when *Vela 3* and *4* picked up a brief flood of gamma rays. Then from 1969 to 1972, *Velas 5A, 5B, 6A,* and *6B*, each over 100,000 kilometers above Earth, detected 16 more mysterious gamma ray events.

Once these bursts were determined not to signify any terrestrial misdeeds, three non-astronomers, Ray Klebesadel, Ian Strong, and Roy Olson, reported their findings in the prestigious *Astrophysical*

Journal. In this short paper, the Los Alamos National Laboratory scientists described brief bursts—some lasting as little as a second—of energetic photons that seem to have originated in space. How far out into space, they couldn't say. At least a million kilometers. Or maybe a few million trillion kilometers, or millions of light-years. There was no way of telling just yet, but everyone knew that if the gamma rays had come from another galaxy, the source would have to be insanely energetic.

Sound familiar?

Thankfully, researchers were willing by this point to entertain such a notion, and they even went looking for coincident reports of supernovae, to no avail. There was as yet no plausible explanation for such beasts.

— Going to Extremes —

Pulsars were proof that something as bizarre as neutron stars existed not just on paper, but in the universe. The pulsar is a laboratory of extreme physics, and its discovery opened up an entirely new field of study. The find was so significant that just seven years later, Hewish shared the Nobel Prize in Physics "for his decisive role in the discovery of pulsars." The exclusion of Bell Burnell, who had treated even the smallest, most unusual bit of data as significant, has been hotly debated ever since.

So what exactly are these objects that have captured the attention of astronomers, watchmakers, car companies, and album art designers? They are nothing like anything you will ever encounter. For that matter, they are nothing like anything you can even really imagine. The textbook description of a rapidly rotating, highly magnetized, city-sized collapsed core of a dead star that began life with a mass between 9 and 25 times the mass of the Sun utterly fails to convey the extreme nature of these objects.

The mass of every living person—over 7 billion of us—on Earth is somewhere in the neighborhood of half a trillion kilograms. Imagine compressing everyone into a thimble. Now imagine something that's a million thimbles wide and a million thimbles tall and a million thimbles deep, each thimble full of similarly compressed material. It's the human population of more than a million trillion Earths inside a ball only 20 or so kilometers across. The extreme gravity of such a mass won't let it be any shape but a sphere, and any imperfections more than about a millimeter tall are crushed by their own weight. On a neutron star, climbing a mountain the size of a rice grain would be more challenging than tackling Everest.

At these densities, material doesn't behave in any intuitive way, but with the help of computer models, astronomers have determined that, like Earth, neutron stars have layers. There is an atmosphere of light elements like hydrogen, helium, and carbon, but this is only an "atmosphere" in the sense that it is the least dense, outermost layer. The entire atmosphere might be a mere ten centimeters (four inches) thick, a scorching million-degree shell with the density of rock. Then there is the outer crust, the neutron star counterpart to what Earthlings walk around on. A bit over a kilometer thick, it consists of the nuclei of iron and other elements along with electrons in a dense crystalline structure that is, at its fluffiest, outermost reaches, about a thousand times as dense as lead. As you burrow inward, things get still more bizarre. Gnocchi, spaghetti, and lasagna appear.

Nuclear pasta—compelling evidence that scientists are frequently hungry while theorizing about bizarre states of matter—refers to the shapes of different, smaller structures within the neutron star itself. These structures are driven by the immense pressures on subatomic particles, most of which are trying to maintain their identities as long as possible. Nuclear gnocchi, for instance, refers to microscopic, nearly spherical conglomerations of neutrons and protons that, if such conglomerations were out in the open, would be called atomic nuclei. But they're not out in the open. If they were, these particular combinations of protons and neutrons couldn't exist. But since the nuclear gnocchi are swimming in a substance that has a density akin to the entire human race per thimble, the usual rules don't apply.

Deeper into the star, the compression increases, and astronomers theorize that the gnocchi are stretched into more spaghetti-like shapes, which coalesce deeper still into sheets like lasagna noodles. At the innermost core, the material is compressed well beyond the density found in an atomic nucleus, and it's anyone's guess what the exact conditions are. Astronomers have been employing an assortment of observational and theoretical strategies for years to crack this particular nut.

What has been agreed upon for decades is that part of a neutron star's interior is a superfluid, the matter flowing effortlessly without friction. This situation is the exact opposite of what one might expect of such a dense material, but one doesn't expect pasta shapes either.

Were a neutron star a completely stationary, nonmagnetic object, things would be much simpler but far less interesting. When a massive star's core first collapses into a neutron star, the neutron star inherits some of the core's original properties. The first is something called "angular momentum." To understand why this matters, imagine a figure skater drawing in her arms. When the mass of her arms is brought closer to her axis of rotation, the ice skater whirls ever faster. Now imagine a figure skater who is not simply drawing a few kilograms of material a few tens of centimeters closer to her center, but who is pulling the mass of half a million Earths into a ball that is a fraction of 1% of the radius that it used to be.

The resulting rotation period makes Earth's day-long one look like an eternity. It's not unusual for pulsars to spin once per second. The one inside the Crab Nebula rotates over 30 times a second. It's exceedingly difficult to put the brakes on this much concentrated mass rotating this rapidly, so the typical pulsar's rotation period is nearly rock steady. Follow-up measurements of Bell Burnell's first pulsar indicated that its timing is accurate to a billionth of a second per year, even more precise than an atomic clock.

We cannot simply spot one of these visually and watch as it rotates, and that's where their extreme magnetism comes in. Just as the original core hands down its rotational momentum to the resulting neutron star, it also bequeaths its magnetic field strength. When the collapse cuts the size of the core by a factor of a million, the magnetic field strength becomes concentrated, but not by just a factor of a million. Instead, the magnetic field becomes stronger by a million *squared*, or a factor of a trillion.

The resulting mega-magnetism, hinted at just a few years prior to the discovery of the pulsar, whips up the charged particles in the vicinity for the same reason that the torrent of charges in the pulsar itself creates a magnetic field. Electricity and magnetism are two sides of the same coin: electromagnetism.

Subjected to a magnetic field of this magnitude, charged particles are stripped from the atmosphere, such as it is, of the neutron star. Along with the protons and electrons that are still swarming among the supernova ejecta, they whirl along magnetic field lines like hyper-caffeinated moths around a streetlamp. Drawn to the magnetic poles, where the field is strongest, these spiraling charged particles emit intense beams of light (electromagnetic radiation) in every imaginable wavelength. If the magnetic poles are offset from the rotational poles— imagine an ice skater twirling while holding flashlights—these beams can become a sort of lighthouse beacon, sweeping across the universe as the neutron star spins. If Earth happens to lie in the beam's path, we catch a quick blip with every rotation. Because the rotation rate is so magnificently steady, that cosmic heartbeat may be the best timing device in the universe.

— Putting on the Brakes —

Something scientists understood well nearly from the beginning of pulsar research in 1967 was this: No matter how smooth the ice, a spinning ice skater will gradually slow, the energy of the rotation sapped by air resistance and friction. The same is true for a spinning neutron star, and within months of the original little green men dis-covery, astronomers had found that although pulsars are astonish-ingly good clocks, they do lose time gradually. This wasn't a surprising discovery because perpetual motion machines, even cosmic ones, are impossible.

As the pulsar whirls around and its magnetic field jostles to and fro while wrestling with the charged particles, it continually radiates away energy. In some cases, a pulsar can emit as much energy as 100,000 Suns, most of which is neither visible light nor radio waves, but the more potent X-rays and gamma rays.

100,000 Suns.

Not all pulsars radiate away that much energy, however. The range of luminosities is frankly enormous, with some pulsars emitting only the energy of one hundred-thousandth of a Sun. But even this figure is incredible, given that you could fit a million billion pulsars into the volume of our parent star.

What boggles the mind is not so much the magnitude of the energy radiated away, which is impressive, but the fact that losing this much energy makes so little difference to the pulsar's rotation. The pulsar in the heart of the Crab Nebula, for instance, continuously sheds as much energy as the Sun, and yet losing this much energy changes its rotation period by only about ten millionths of a second per year.

For most pulsars most of the time, spin-down is a smooth, gradual process, which is what astronomers expected once they settled on the mechanism behind the pulsar itself. Within just 15 months of Bell Burnell's initial discovery, 35 more pulsars had been spotted, 19 of which were observed by Australian radio astronomers, who were just warming up.

Things were falling nicely into place as far as mapping and timing went. Pulsars were being found in a part of the Galaxy known as the "disk," a vibrant, spiral-armed pancake of activity where new stars are being born and short-lived stars are dying to make neutron stars. In contrast, precisely zero pulsars had then been discovered in the galactic retirement homes known as "globular clusters." These ancient spherical swarms of stars are situated above and below the disk of the Milky Way, and for the most part, they are populated by the oldest stars in the Galaxy. No doubt, massive stars blew up deep in the past, but over the last few billion years, the ice skater's energy would have been completely sapped. Astronomers agreed that neutron stars would no longer be pulsars in those environments.

Meanwhile, they were finding out other pulsar "truths." For instance, pulsars with the strongest magnetic fields were the ones most likely to be found inside obvious supernova remnants (e.g., the Crab Nebula), and because they had such strong magnetic braking systems, they were also the ones that were slowing down the most. Researchers dutifully plotted quantities like pulsar slowing rate against pulsar period, magnetic field strength, and age, and they saw definite trends.

It all made sense—for a city-sized ball of supermagnetic material denser than a mountain per teaspoon.

CHAPTER 7

Stellar Arrhythmia

Curiouser and curiouser.

—Lewis Carroll, *Alice's Adventures in Wonderland*

— Clockwork Classroom —

It was shaping up to be a hot, sunny February day, and a few dozen students from one of the local high schools began filing into the Science Operation Centre at Australia's Commonwealth Scientific and Industrial Research Organisation (CSIRO) Space and Astronomy facility in a suburb of Sydney. Each decked out in a burgundy and gray school uniform and sporting a "Visitor" lanyard, the teens awkwardly took in their surroundings. Two offices with banks of computer monitors were on their left. A bowl of fruit and cereal bars was parked on the middle table (those were snacks for the astronomers observing overnight). And to their right were portable dry erase boards. A large smart screen sat front and center in the room, and on it was a live image of Murriyang, the Parkes Radio Telescope. A few wispy clouds swept by the Dish, but otherwise it was a similarly hot, sunny morning at the telescope, which was a five-hour drive away. But it wouldn't have mattered what the weather was. This radio telescope can operate 24/7, through clouds, during the day, any time. It's one of the perks of radio astronomy.

On that day, the high school students would be deciding on the observing targets for the Dish.

Education manager Robert Hollow gave a quick slide presentation and laid out the basics of pulsars. They are incredibly dense, rapidly rotating neutron stars that beam intense radiation into space like a lighthouse, he explained. And the radio telescope they were about to use had single-dishedly discovered over half of the nearly 3,000 pulsars then known. The students got a quick rundown of the celestial coordinate system and the timekeeping conventions of astronomers, along with a primer on figuring out where in the sky a pulsar might be at any given moment. Then they were given a list of 47 representative pulsars and their coordinates—at which point one student declared that pulsar names like PSR J0437-4715 were "boring"—and they were off. In small groups and with all of perhaps an hour of training, they had to decide which sequences of letters and numbers seemed most intriguing and, more important, most readily observed that morning. Then they told the 64-meter-wide radio eye where it should look next.

The Dish began to slew to the first target amid oohs and aahs, and within minutes the first pulsar's data appeared, along with a steady stream of questions. "What does this diagonal line mean?" asked one student, looking at the third graph plotted on the computer screen. Like a Halloween-themed backslash, a glowing orange stripe cut downward and to the right on the plot's dark background. Hollow explained that the graph depicts the arrival time of the different frequencies.

The telescope has detectors that can pick up not just a single radio wavelength, but a vast range. The lowest frequency they were observing that day was 1350 megahertz (MHz), corresponding to a wavelength of about 22 centimeters (nearly 9 inches), and the highest was nearly 1500 MHz, with a wavelength of about 20 centimeters (about 8 inches).* The students' plot seemed to indicate that the higher-frequency, shorter-wavelength light, which was at the top of the graph, had shown up just a little bit earlier than the lower-frequency, longer-wavelength light for each pulse of the pulsar.

"But doesn't all light travel at the same speed?" asked another student.

Hollow explained that, yes, in a vacuum, all light travels at the same speed, even radio waves. But interstellar space isn't a vacuum. There is the thinnest fog of atoms, protons, and electrons. In this fog, you might find as little as one stray electron or as much as a few million atoms and other particles in every cubic meter. The latter figure

---

* Every second, approximately 1.5 *billion* radio waves with a wavelength of 20 centimeters zip past you.

might sound impressive, until you compare it to the 10 trillion trillion atoms found in a cubic meter of sea-level air.* But those few million particles per cubic meter add up for a light wave traveling hundreds of trillions of kilometers from pulsar to telescope, and as the radio waves travel through the thin fog, the lower frequencies find themselves stopping to shake hands with the charged particles they encounter. The more distant the pulsar or other radio source, the longer the delay astronomers see between the arrivals of the higher-frequency light and the lower-frequency light.

A lightbulb appeared over the student's head. "So this would be more slanted if it were farther away?"

Exactly.

I marveled at how quickly these untrained astronomers-for-the-day were grasping the implications of the data that appeared in front of them. Pulsar astronomers indeed use the slope of the frequency-phase graph to help gauge the distances to the pulsars they observe. This "dispersion measure" is a fundamental quantity that features in pulsar catalogs, including the definitive one maintained by the Australia Telescope National Facility, the collective name for the radio observatories managed by CSIRO, including the Dish.

The first observing team switched out, and a new group of students settled in to look at their chosen pulsar. The enormous eye of the Dish rolled eastward, and the data taking commenced. There was some murmuring, and a tentative hand went up.

"I guess we didn't point it at the right place?" was the statement-question.

"Ah, looks like you've got a nulling pulsar," Hollow declared brightly, as if that explained everything.

"So, what does that mean? It isn't pulsing? Why not?"

This question revealed a fundamental naïveté in the student. He expected the astronomers to know the answer to such an obvious question. Surely with over five decades of observing these beasts, astronomers understood them. But astronomers don't know why some pulsars like PSR J1717-4054 suddenly seem to turn off and turn on again, nor

---

* That's 10^{25}, or 10 heptillion.

do they understand why some stay off most of the time, or what factors contribute to the precise profile of each pulse, or . . .

The unanswered questions seem endless.

What the students did not question, at least that time, is how astronomers know that the pulsars on their list exist in the first place. The plots they obtained were nothing like Jocelyn Bell Burnell's end-less streams of paper, each individual pulse showing up as a small hump in an otherwise mostly level landscape. Instead, they saw on one plot a single, tall, narrow spike on that flat landscape. Bell Burnell could have manually done the same thing had she taken a long strip of that first pulsar's readout and folded it into smaller, equal-width strips, each representing 1.3373 seconds of data with each blip squarely in the center. Having done that, she could have added up all the blips to create one tall spike in a much smaller lawn of grassy-looking noise.

The nice thing about random background noise, as it turns out, is that one second it might be a touch higher and the next second it might be a touch lower. A real signal, though, is consistently a touch higher, even if it's sometimes dwarfed by the noise. As a result, when astronomers add up all the data, the background noise doesn't add up nearly as quickly as the reliably stronger signals, which keep growing with each addition.

And thank goodness for that. The signals of most individual pulses barely rise above the noise, so simply eyeballing the data for repetitive blips misses all but the strongest pulsars. Bell Burnell's first pulsar was practically screaming at us, but countless pulsars no doubt passed undetected over her telescope.

As computing power grew in the early 1970s, pulsar researchers turned to something called discrete Fourier transforms (DFTs) to do the detecting for them. A DFT is essentially a computational algo-rithm that can try out every possible folding width to determine if there is a repetitive, regular signal coming from the direction the tele-scope is pointing. It's an incredibly handy tool, but it works only if the pulsar behaves itself, or at least mostly behaves.

— Glitchy —

Early on, it seemed that all the pulsars were on their best behavior. Then, amid all the sensible pulsar findings came a rather jolting shock.

The pulsar associated with the Vela (no relation to the satellite program) supernova remnant had abruptly and without any warning begun spinning faster. Soon after, in 1969, astronomers marveled that the Vela pulsar's rotation after "this extraordinary event, astonishing even by pulsar standards," had proceeded to gradually slow down as though nothing had happened. Rumor has it, it even began whistling innocently.

The immediate and most obvious explanation was that something had caused the neutron star to suddenly shrink. Not by much, mind you. Calculations indicated that such a speed-up in rotation could be caused by the pulsar becoming perhaps one centimeter smaller in radius. That minuscule adjustment would have been enough to speed up this ice skater's rotation rate by two parts in a million, easy enough for the radio telescopes to pick up.

Before long, more such "glitches" were discovered in other pulsars, and astronomers scrambled to figure out why. The answer seemed to be that a glitch is actually a starquake, a stellar Krakatoa where the crust suddenly shifts to relieve a building tension. The tension itself, astronomers reasoned, would arise from the slowing spin. Like the ice skater's skirt, which appears to be flung out as she spins, a pulsar's equator would also be flung out, making it less spherical and more oblate. But as the pulsar slows, this flinging-out strength would gradually diminish, and the pressure on the surface of the neutron star would gradually increase. At some point, just as with a terrestrial Krakatoa, something would have to give. When the tension became too much for the crust to endure, the neutron star would suddenly—in less than a minute—lose some of its oblateness, shrink to an imperceptibly smaller radius, and spin just a tiny bit faster.

Well, it looked good on paper anyway. Neutron stars are, as I have mentioned, nothing like you can even imagine, and the underlying explanation for a glitch quickly became far more involved than imagining figure skaters or Krakatoas. Moreover, any given pulsar would need to adjust its crust in this fashion only once in a few centuries, but some were observed to glitch multiple times.

Still, early observations of pulsar glitches revealed to astronomers one very important thing. A single, brief glitch in a modest pulsar can release the same amount of energy that the Sun releases in a few *de-cades*. Or, if you prefer, it releases a few nano-foes of energy in a hand-ful of seconds. Imagine catching all the sunlight that has been pumped out over the course of your life. That is, give or take, the amount of energy associated with a single pulsar glitch. These are extreme events.

So if it's more complex than a settling crust, what *is* going on? Let's forget the ice skater for the moment. Think of a pulsar as a spin-ning raw egg. You can see the exterior rotating. The interior rotates as well, but it's not fixed to the shell the way it would be if the egg were hard-boiled. A go-to trick for an elementary school magician is to spin an egg rapidly, briefly stop its spin, utter a magic word, and quickly release the egg. Magically and mysteriously, it starts spinning again. The interaction between the still-churning interior and the station-ary shell is what speeds the egg back up, but the audience can't see the inner workings through the shell.

Interactions between eggshells and egg whites are fairly straight-forward, but inside a pulsar, the interplay between what astronomers can observe—the outermost crust—and the inner layers is not so clear. In fact, it sometimes seems to involve magic words, or at the very least immense magnetic fields, superfluid cores, and vortices. The exact dy-namics of glitches are still very much a topic of debate, even with so-phisticated computer modeling. Some astronomers have even gone back to square one, suggesting that smaller starquakes resulting from crust stresses are catalysts for the larger glitches.

All this glitchiness might bring into question whether a watch-maker really should name itself after a pulsar. But glitches are rela-tively rare events, and for the rest of the time, pulsars are exquisite timekeepers. If a regular pulsar isn't good enough for your timing needs, though, it's time to meet some even more extreme denizens of the cosmos.

— Ain't Misbehavin' —

The best-behaved pulsars, at least in terms of their timing abilities, are known as "millisecond pulsars." They are exactly what they sound like: pulsars whipping around on time scales of milliseconds.

Thousandths of a second.

As of this writing, about 200 pulsars are considered millisecond pulsars, and the current record holder is PSR J1748-2446ad, which rotates 716 times per second and will henceforth be called Zippy in this book. The blade of a typical kitchen blender on the highest setting, by comparison, rotates a mere 500 times per second. One can read numbers like this and think, "Oh, that's interesting," but grasping the reality is impossible.

You should still try.

Half a million Earths of matter.

Twenty kilometers across.

A kitchen blender.

A point on the equator of one of these beasts moves at nearly a quarter of the speed of light. Were the pulsar spinning much faster, even its gravitational superglue would not be enough to hold it together. It is physically impossible for a pulsar to spin any faster than about 1,500 times per second and not fly apart, a fact that gave early pulsar astronomers a handy limit to the sorts of signals they should be looking for.

The original millisecond pulsar, PSR B1937+21, was discovered with an enormous radio telescope in Puerto Rico in 1982. The 305-meter-wide Arecibo Observatory was built in a natural sinkhole in 1963, and it was the largest single radio telescope in the world until 2016 when the Five-Hundred-Meter Aperture Spherical Telescope in China claimed the title. Looking like an enormous shallow bowl, or perhaps a skateboarder's dream, dropped into a tropical setting, Arecibo was something of a pop-culture icon during its reign. It featured in 1995's *GoldenEye* as well as the 1997 movie *Contact*. Tragically, the Arecibo radio telescope began showing its age in the late 2010s and was irreparably damaged by a series of catastrophic cable breaks in 2020 that sent the support towers crashing through the dish itself. It was a devastating gut punch to the astronomical community, the sudden death of an old, trusted friend.

The extensive list of accomplishments made possible by Arecibo could fill volumes. It was particularly valuable to pulsar astronomers, who were the first to discover planets outside our solar system and the first to verify certain aspects of Einstein's general theory of relativity.

Finding something as extreme as PSR B1937+21, the first millisecond pulsar, was just one of many testaments to the supremacy of this telescope. Just a few degrees away from Bell Burnell's original pulsar, PSR B1919+21 (hence their similar names), this enigma rotated once every 0.001557708 seconds, its period steady to one part in a quadrillion.

But wait—neither of those figures made sense. Backer and colleagues, the authors of the initial article in *Nature*, were essentially at a loss, stating, "The existence of such a fast pulsar with no evidence either of a new formation event or of present energy losses raises new questions about the origin and evolution of pulsars."

At least they didn't say it was "out of the question."

The first millisecond pulsar challenged practically everything astronomers understood about pulsar rotation. Its dispersion measure placed it about 12,000 light-years away, handily inside the Milky Way Galaxy. Pulsars slow as they age, particularly at the earliest part of their lives. That meant that if this one were spinning so quickly, it must be brand-new. Right? Initial estimates suggested that it had to be younger than the Crab Supernova of 1054, but if that were the case, surely there would have been historical records of it. And even if there weren't records for some reason, surely the pulsar would still be buried in the explosive mess from the progenitor star's supernova. Right? Radio, X-ray, and optical searches turned up nothing. And in any event, with such a high spin rate, it should be slowing down dramatically as its magnetic field acted as a cosmic dynamo. Right?

So how could its rotation rate be so steady?

Pulsar astronomers scratched their collective heads as they looked at the clear pulse profile of big blip, smaller blip, big blip, smaller blip. Finally, the clue to their origins came not from the pulses, but from the company they kept. Millisecond pulsars were being discovered disproportionately in globular clusters, the galactic retirement homes where almost no "normal" pulsars reside.

What's more, most weren't acting alone.

CHAPTER 8

(Almost) No Star Is an Island

Imagination is the only weapon in the war with reality.
—LEWIS CARROLL, *ALICE'S ADVENTURES IN WONDERLAND*

— Where's the Kaboom? —

Astronomers eagerly anticipated the brief shining moment from an object known as KIC 9832227, one of 13 million possible observing targets in the expansive Kepler Input Catalog. Rather than a single object, KIC 9832227 is actually two stars sharing a common outer atmosphere, and if calculations published in 2017 and reported worldwide had been correct, the two would have gradually merged into a single star and flared up as a fiery red nova in the year 2022. It wouldn't have been nearly as spectacular as SN 1006, but it would have boosted this otherwise invisible dot to something easy enough to spot in the suburbs. Unfortunately for those of us waiting for even a modest visual transient in our sky, further scrutiny of the system showed that while it might eventually make a cosmic spectacle of itself, it won't do so for quite a while.

Bummer.

The two culprits hiding behind the curtain in KIC 9832227 are, by themselves, rather unassuming stars. One weighs about 1.4 times the mass of our Sun, and the other is a runt, a mere 30% of the Sun's heft. Together, they don't create a star even twice as massive as the Sun, so there won't be any core-collapse supernovae or pulsars in this pair's future. Separated by just 2 million kilometers—by contrast, Earth is 150 million kilometers from the Sun—both objects are swimming around in the same extended gaseous envelope, a bit like a large goldfish and a small goldfish doing laps inside a fishbowl.

Eleven-hour-long laps inside a very big, very hot, and somewhat dumbbell-shaped fishbowl.

The 2017 measurements seemed to indicate that the laps were getting progressively smaller and progressively quicker, and at some point very soon, the two stars would collide. The collision was expected to pump out the light of a million Suns for a brief time, fading away over the course of a few weeks, and leaving behind a single red supergiant star with no further hints that it began life as two stars. From there, it would have died the relatively uneventful death of a star with nearly twice the mass of the Sun, following the progression on the poster found in astronomy classrooms worldwide: The giant would slowly slough off its outer layers, reveal a planetary nebula that would eventually disperse, and leave behind its exposed white dwarf of a core. Still, it would have had its moment of glory, albeit not as glorious as a supernova, but a more impressive phase than the Sun will ever manage.

Unfortunately, due to computational errors in the 2017 research, KIC 9832227's 15 minutes of fame will have to wait. Headlines went from the breathless "Exploding Binary Stars Will Light Up the Sky in 2022" to the dejected "Two Stars Won't Collide into a Red Nova in 2022 after All (We're Disappointed, Too)."

For some readers, the takeaway from this episode was a skepticism about the usefulness of science, an endeavor that always seems to be changing its mind. Instead, I see this incident as a beautiful illustration of how science works. Science team A reports something, and science teams B, C, D, and E double-check those findings and incorporate new data and new methods of analysis. That double- and triple-checking is particularly important if the original study promises something as cool as a naked-eye transient. In this case, it didn't take long for subsequent studies to spot the error, and with the cooperation of the original team, the prognosis was revised.

If you think about it, KIC 9832227 should be anything but a disappointment. It is effectively two stars in one, which raises a number of questions: Is this normal? Do stars often find themselves sharing a stellar fishbowl with other stars? How do they get to that point? And how do astronomers spot a two-in-one star in the first place?

— Fraternal Twins (and Triplets and Quadruplets) —

The Sun is an oddball. With its collection of minuscule planetary companions, it treks lazily—only about 250 kilometers per second—around the center of the Milky Way Galaxy. It has no comparably sized stellar partner sharing its journey, and there's no real evidence that it ever did.

But most stars do. The eminent astronomer Cecilia Payne-Gaposchkin was reportedly known to say, "Three out of every two stars are in a binary system." This seems at first to be a bit of a riddle, but it's just a roundabout way of saying that half the dots you see when you look up at night are, in fact, not just single dots. Indeed, it's likely that fully 80% of the stars in the universe are members of multiple star systems. Binary systems—those with two stars—are the most common by far, but hiding in plain sight in the constellation Gemini the Twins is a six-star system. Thought for millennia to be a single star, Castor was found in the 1800s to be, at the very least, a binary star. As technology improved, it was eventually discovered to be three binary systems consisting of two stars hotter, brighter, and more massive than the Sun and four cooler, dimmer, and less massive stars. In the case of Castor, there are six dots for the one you see.

Finding a sextuplet system is always a treat for astronomers, but they're not completely unexpected. Stars are born in batches in enormous clouds of gas and dust capable of producing hundreds or thousands of stars at a time. In such close quarters, being paired with a stellar dance partner at birth is commonplace. With the help of supercomputer simulations, astronomers have been able to virtually speed up the process of star formation from something spanning millions of years to something the duration of a commercial break. In simulations like STARFORGE (*STAR FOR*mation in *G*aseous *E*nvironments), you can watch as a diffuse cloud with 20,000 Suns of matter condenses into filaments, which become beaded with hot dense knots that then turn into stars.

If we could get a close-up of these seemingly tiny knots, we would see that they are enormous whirling disks of material millions of kilometers across. As matter from the gas cloud tries to gather on the forming star, it moves ever faster, subjected to the same laws that spin ice skaters and pulsars. The accreting material flattens into a disk, which may then fragment into another knot of material with its own disk. Now there are two stars-to-be locked in a gravitational dance with each other.

All of this plays out over an unfathomably long time, but it doesn't take place in isolation. Amid the ongoing star formation are what appear to be cosmic blowtorches—swirls and whorls of hot gas spewing from dying stars—and supernova explosions of the most-massive, shortest-lived stars. Sped up to a pace that humans can comprehend, star birth reveals itself to be a dynamic mess.

But in that dynamic mess, one thing becomes evident. Even when the unused gas and dust are finally dispersed, most of the newborn stars remain in a relatively tight clump (while some get ejected entirely), and most of them have at least one close fraternal twin. Whether or not a star is born with a companion is strongly dependent on how massive the star is, and that depends on some of the more exact details of its birthplace. Higher-mass stars are more likely to be in a binary system than are lower-mass stars, and it's much harder to find a star with, say, 20 times the mass of the Sun hanging out by itself than it is to find such a star in a binary pair.

Despite their prevalence, binary systems are not always obvious, and it's the odd couple that presents itself as two separate dots in our telescopes. In fact, the telescope had been around for over 150 years before the first visual binary was definitively announced. In a binary system, one star doesn't orbit the other, but instead they both orbit a common center of mass, a balance point that lies closer to the more massive star than to the less massive one, just as the adult on a seesaw sits closer to the fulcrum than the child does.

If astronomers are lucky, they can spot both stars orbiting this point as the days, months, and years progress. Most likely, though, the presence of a companion has to be divined through sneakier channels. There are binary systems that are revealed only through their spectra. As one member of the system heads away from us, the other necessarily heads toward us while emitting the Doppler-shifted "eeeeeeeeeeeeeee" to the first star's "yoooooooooooooooooo." Except, of course, they are emitting light waves, not sound waves. As one star moves to the left, the other moves to the right. Such is the nature of the binary dance. Here on Earth, astronomers pick up those subtle, regular changes to the wavelengths in the stellar bar codes, and with a bit of analysis, they can pick out the finer details about the system.

There are some systems that we are lucky enough to observe from the side. In these special cases, the stars take turns passing in front of each other as they orbit, each blocking out at least some of the light of their partner star once every orbit. These eclipsing binary systems show up as objects that periodically dim, the exact rhythm and extent of that dimming dictated by the masses, temperatures, and separation of the stars.

— Kissing Cousins —

Mutual eclipsing is how astronomers first grasped the binary nature of KIC 9832227. Launched in 2009, the Kepler Space Telescope had been keeping its eye on this target from the beginning. To be fair, it had also kept its eye on over 160,000 objects 24/7 for the entire four-year duration of its original mission, and then on hundreds of thousands more for the remaining five years that it was operable. Finding stars whose light output varied was the Kepler telescope's forte and its raison d'être. It used this method to discover minuscule eclipses—known as "transits"—as distant stars were partially blocked by their own planets.

It was an astonishingly simple plan. Watch a bunch of stars. A whole bunch. See if any of them get dimmer and brighter with a regular rhythm. Flag those for follow-up observations. Over its mission life, Kepler discovered literally thousands of planets orbiting other stars. But it also gave astronomers an unprecedented view of pretty much everything else that could make a star's light vary, from pulsating stars to stellar flares to mutually eclipsing stars to supernovae. Astronomers realized that KIC 9832227 was behaving like something called a "common envelope binary," a special system where two stars are essentially roommates living amid their shared possessions.

This is not a typical binary star fate. Some binaries are in truly long-distance relationships, the two stars separated by as much as a light-year or more and barely aware of each other's existence. Such widely separated stars move casually around their balance point, some pairs taking thousands of years—even millions—to make a full circuit. When stars are so tenuously connected, it takes just the smallest gravitational nudge from a third passing star to break the bond. Meanwhile, in the closest binary systems, the two members can whip around each other in what might seem an impossibly short time of mere days or hours—or even minutes. This tarantella makes the 88-day-long orbit of the planet named for the winged messenger Mercury, which orbits the Sun at a distance of 63 million kilometers, look like snail mail. Interestingly, it's often a third star that puts them in this position, nudging the pair together instead of breaking their connection.

Unless the stars are relatively close—within a billion or so kilometers of each other—life progresses as though they are each solo stars. But having a less standoffish partner can dramatically alter a star's career path. Instead of succumbing to the fate charted by the poster in an Astronomy 101 classroom (hydrogen fusion to giant to planetary nebula to white dwarf. The end), a partnered star with less than about nine times the mass of the Sun has options. "The end" . . . isn't. At least not necessarily. While it's true that the first phase of hydrogen fusion carries on as though nobody else is around, things can get far more interesting once that phase is over.

For a star like the Sun, that first phase lasts 10 billion years, and for lower-mass stars, it lasts even longer. Given that the age of the universe is only 13.7 billion years, considering the ultimate fate of the lowest-mass stars is seen by some as a pointless exercise. While extremely long, 10 billion years is not forever. Ultimately the star runs out of fuel in its core, and gravity takes charge once again, shrinking and heating the core and releasing that energy to the outside of the star, which responds by cooling and expanding.

At this point, the sharing can begin. Each star has a large teardrop-shaped gravitational territory known as its "Roche lobe," the odd shape arising because the two stars aren't simply sitting there in space but are in a mutual orbit around their balance point. There is a spot between the orbiting stars where a particle feels just as much gravitational attraction to one star as it does to the other. This spot is the tiny nexus where the points of the two teardrops meet, and it's the doorway through which matter from one star can enter the territory of the other if it gets a little too big for its allotted space. As the more massive primary star's outer layers bloat, it begins to unapologetically spill its material into the personal space of its companion, which, because it has a smaller mass, is living life at a slower pace.

Losing mass through this doorway has some profound effects on the primary star. First, its own teardrop shrinks. The size of the Roche lobe is tied to the mass of the star, so if it loses mass, it loses some of its gravitational territory. But when its Roche lobe shrinks, it needs to lose even more mass to stay within its teardrop, so it loses mass even faster as its teardrop territory closes in. Sometimes, the process succeeds in stripping off much of the star's outer layer, exposing its hot, dense, not-quite-a-white-dwarf-but-kind-of core, and sometimes the star manages to shrink away from the encroaching walls and kick-start a new life as a slightly less massive star. Meanwhile, the companion is typically too polite to refuse what is being given to it, and so its mass and Roche lobe grow. It then can proceed to live a slightly faster life as a slightly more massive star.

But sometimes the companion puts its metaphorical foot down. Dumping mass from one star to another is nothing like filling a bucket with a garden hose, and as the seething hot gas pours inward toward its prospective new owner, the entire dynamic of the system is altered. Material swirls in a disk around the receiving star, heating as it edges closer. The star swells in response to the rise in temperature, and depending on how everything is arranged, it can fill its own Roche lobe. Now the hot gas from two swelling stars spills out beyond either's gravitational territory, leaving the stars inside a common envelope, which belongs to neither star but affects both.

In the case of something like KIC 9832227, the new arrangement is a real drag. Friction from the common envelope saps energy from the orbiting stars, causing them to move in ever-tighter orbits. Because this energy is being pulled out of the orbits and deposited into the common envelope, the fishbowl in which these two goldfish are swimming grows ever larger and more diffuse. At some point in the extremely distant future, the shed envelope of this system might make for a pretty, double-lobed, glowing planetary nebula. Sporting all sorts of interesting structures that betray the stars' earlier relationship, this nebula will emerge long after the innards of the two original stars merge into one and give Earthlings a view of a luminous red nova.

But it didn't happen in 2022.

— In the Windmills of the Hard Drives —

The KIC 9832227 miscalculation is completely forgivable. After all, multiple star systems are a messy business. They're hard to observe, and they're nigh impossible to model, as theorist Orsola De Marco knows all too well. In a universe where so many stars have had a life-changing run-in with at least one other star, it makes sense that those interactions will leave an imprint on what we see. The problem is that there are so many possible ways for stars to interact and finite resources to compute the outcomes.

"Ideally, we want a three-dimensional model that has incredibly high resolution both in time and space, and you need a large volume to model the whole orbit of your binary, but you also want to be able to model the tiny scale of the stellar core, which is [the] size of the Earth." She continued to tell me her wish list. "I want to model the mass transfer that starts when the star expands. I want to model hundreds of thousands of years with all the physical laws. I want magnetic fields. I want full general relativistic effects. I want the whole thing, and I want it right now."

A song from the 1971 movie *Willy Wonka and the Chocolate Factory* ran through my head.

But astronomers can't get everything they want, so models necessarily cut some corners. De Marco's models give more detail about what's happening on the smaller scales than STARFORGE does, but they cover less time.

Once acceptable compromises have been made, astronomers send all the physics of their interacting stars to a supercomputer and have it churn through the stars' future. When I talked to her, De Marco's latest results from the Gadi supercomputer had just come in. They had taken six months and several hundred thousand dollars to complete.

I gasped. Gadi is *the* supercomputer in Australia, the very state of the art. It is the fastest supercomputer in the Southern Hemisphere, and the computations had still taken six months.

"But that's okay," she assured me. "A simulation like that is very valuable. You run simulations, and then others can mine them for information."

True. Simulations like De Marco's, STARFORGE, and others become pieces of a virtual universe. Astronomers interested in one subfield might explore that aspect while other astronomers can explore an entirely different aspect of the exact same simulation. In fact, it's rare that a simulation is just one astronomer's computational baby. The input and output, along with the cost, are frequently shared among researchers around the world.

"Do you want to see one of our simulations?" she asked.

Of course I did.

She tapped a few keys and brought up a video. "This one is by my student Miguel Gonzalez."

Soon, there was a stellar dance playing on the computer, along with Noel Harrison crooning "The Windmills of Your Mind."

"Round, like a circle in a spiral . . ."

"I put it to music because I thought it was better that way," De Marco explained.

It was.

As the song played, I watched a gossamer yin-yang with two hot dense patches. The eyes swirled around each other, and streamers of hot gas shed into space.

"Never ending or beginning on an ever-spinning reel . . ."

The simulation paused to give the observer the chance to look at the system from above or the side or even to zoom in.

"And the world is like an apple whirling silently in space / Like the circles that you find in the windmills of your mind."

As a work of animated artwork, this video paled next to what Pixar's artists can achieve. But a computer simulation is so much more than a movie. Each point carries information about the temperature, velocity, density, or other properties, and from those, astronomers can figure out what happens next in the story. All in all, the three-minute simulation told the tale of just 20 years in the life of a binary system, a time span that is, cosmically speaking, no time at all.

The baton is now passed to another research group keen to explore the next stage in the evolution of the system, and the simulation is recycled and repurposed to answer new questions.

— Recycling Program —

This naturally brings us back to Zippy, also called PSR J1748-2446ad, also called Ter5ad, the fastest-spinning pulsar known. At 716 rotations per second, its sonified signal sounds like a note 1.5 octaves above middle C, and there is no known mechanism for a single massive star to result in such a beast. Zippy is not a single star. Its companion is an extremely puffed-up ball of hot gas toting less than one-fifth the mass of the Sun in a package five times wider. Zippy outmasses its partner by a factor of ten, but its matter is compressed to the size of a city's downtown area. The two objects orbit their gravitational balance point in just over one Earth day as they hang out in a dense spherical star cluster known as Terzan 5.

Lying 18,000 light-years away in the general direction of the constellation Sagittarius, Terzan 5 is an unusual system, most likely the remnant core of a galaxy largely consumed by the Milky Way billions of years ago. It is estimated to have contained as many as 100 million stars in its heyday, but now there are only 2 million tightly huddled stars, including three dozen of the 230 millisecond pulsars known by 2022. Both of these are impressive figures given that the cluster is only five light-years wide. Admittedly, five light-years is unfathomably large for creatures that find cross-country distances enormous, but for comparison's sake, a five-light-year-wide ball centered on the Sun would contain the Sun.

That's it.

To encompass the Rigil Kentaurus system, which consists of three stars, our imaginary sphere would have to be nearly nine light-years wide.

The core of Terzan 5 is so packed with stars that you would be hard-pressed to know whether it's day or night. In such an environment, interactions between stars are common. Even if they weren't born together, stars can quickly find themselves gravitationally herded into close binary star systems where the evolution of one star has a profound impact on another.

Something certainly had a profound impact on Zippy/Ter5ad/ PSR J1748-2446ad and all its millisecond pulsar brethren. Astronomers had over 20 years of millisecond pulsar observations under their belts by the time Zippy was discovered. Like most of these oddities, Zippy was found to be part of a binary system, strongly suggesting that a partner was to blame for getting the pulsars so wound up. But how?

The situation begins with the fast life and explosive death of a massive star, but this time it has a companion living life at a slower pace. Time passes, and the second star ages while the original pulsar inevitably slows down, possibly even outliving its mega-magnetic pulsar phase and becoming a relatively dead, slowly spinning neutron star. But the partner breathes new life into it, spilling itself onto the neutron star and creating something called a "recycled pulsar." From there, it's the ice skater in action again. As matter is drawn toward the center of rotation, the ice skater rotates faster. And faster. And faster. As long as something is continually spiraling toward the axis of rotation, the ice skater will continually speed up.

Now imagine that the skater is a massive, compact, magnetic beast that might kill its partner when it finishes eating.

— When Stars Become Spiders —

Dropping matter onto a pulsar is an ill-advised but irresistible pursuit for its partner. In the case of the Transformer pulsar, officially designated J1023-+0038, it will ultimately prove to be fatal. Falling shy of Zippy's 716 rotations per second, Transformer still sports an impressive 600 rotations per second.

Transformer was originally misidentified in 2001 as a system containing an unusually magnetic white dwarf and a low-mass companion, but later observations revealed that the magnetic white dwarf was, in fact, a much denser pulsar that didn't always behave like one. Sometimes it pulsed. Sometimes it didn't. What intrigued astronomers was not so much that it occasionally nulled—lots of pulsars do that—but as radio telescopes lost its signal, X-ray and gamma ray telescopes picked it up.

Plenty of pulsars had been detected with orbiting X-ray and gamma ray observatories before, but this bizarre one was continually switching identities. In 2013, after carefully observing it with the Fermi Gamma-Ray Space Telescope and an assortment of ground-based radio telescopes, astronomers finally worked out a scenario that made sense.

Transformer is a black widow.

During its radio pulsar phase, it is a fairly typical pulsar, which is to say it is a tornado inflicting devastation on everything around it. To be around a pulsar is to feel a blast of high-speed charged particles. Its companion, a puffy red star with barely a fifth of the mass of the Sun, feels that blast keenly as it orbits once every 4.8 hours in a tight circuit around Transformer. As long as Transformer is on, its strong wind keeps the material from its partner at bay. But when it turns off, hot gas from its companion enters its gravitational territory and races ever faster around the pulsar. As it spirals toward this gravitational menace, the ultrahot disk spews out the most energetic light possible: gamma rays and X-rays. Then, for reasons known only to Transformer, the pulsar kicks back on again as a radio pulsar. Its intense wind blasts away what its partner dumped on it, and the cycle starts anew.

After enough of these identity crises, nothing will remain of Transformer's partner. It will have been completely eroded by the pulsar's alternating acceptance and rejection of its offerings, eaten alive by its black widow mate.

It's a depressing fate for the companion, to be sure, but it has been a fascinating case study for astronomers. And if there's anything the universe does remarkably well, it's cook up fascinating systems.

CHAPTER 9

The Making of a Superstar

The report of my death was an exaggeration.

—Mark Twain, in an interview with Frank Marshall
White for the *New York Journal*

— Impostor Syndrome —

I caught Nathan Smith as he was in the midst of observing a cosmic enigma that has defied explanation for nearly two centuries. He was sitting comfortably at his home in Arizona while remotely operating the Baade Magellan Telescope perched over two kilometers high in the Atacama Desert in Chile. Named after Walter Baade, this 6.5-meter telescope shares the site with its twin, the Clay Magellan Telescope, and other telescopes at Las Campanas Observatory. While Smith was observing for the night, I was 15 time zones ahead with a million questions about something that had intrigued me for decades: Eta Carinae.*

It had obviously intrigued Smith as well, given that he has authored several dozen papers on the subject. During our exchange, he revealed that he was never a huge fan of science as a teenager, preferring skateboarding and music instead. But, he admitted, volcanoes seemed cool. When he took an astronomy class as an undergraduate and found out that Mars had the biggest volcano in the solar system, he sat up and took notice.

"Then we started talking about stars, and then massive stars, and eventually supernovae," he recalled. "And I realized that planets are super boring because they don't explode."

---

* In a sampling of YouTube videos, I have heard various astronomers discuss AY-ta car-EYE-nay, AY-ta CARE-in-ay, AY-ta car-EE-na, and even EE-ta car-EE-nee. But its variable pronunciation is one of the least puzzling aspects of this system.

But neither has Eta Carinae.

Not completely, anyway.

Looking at Eta Carinae is a bit like taking an inkblot test. To one person, it looks like an impressionist's carnation on a still, glassy lake. To another, it's an hourglass made of two mushroom clouds joined at the base. A Hubble Space Telescope image hanging on my office wall reveals what looks to be a billowing, glowing blast from a generic action movie, along with its mirror image. But the movie is frozen, as though the universe's processors are too overwhelmed to render the rest of the explosion.

A time-lapse animation of the dual-lobed, roiling gas clouds shows a clear expansion over a quarter of a century, but the main structures—a dark I-shaped patch here, a bright figure eight there—remain. As far as stellar detonations go, this one appears pretty noncommittal. In fact, Eta Carinae has been called both a failed supernova and a supernova impostor, pretty harsh insults for an object that in 1962 was the subject of a love letter by astronomer A. David Thackeray: "Among remarkable objects in the Southern sky, surely the most remarkable of all is Eta Carinae."

On the other hand, astronomer Chris Sterken wrote that "it simply is the photometrist's ultimate nightmare." Whether it's remarkable or nightmarish—or both—Eta Carinae is undeniably an odd bird that has captured humankind's attention before and will certainly do so again in the most spectacular fashion. In the meantime, astronomers like Smith are finally making some headway in understanding how it got into its current mess.

To the naked eye, Eta Carinae—or more appropriately η Carinae—is a modest-looking, reddish dot in an overcrowded patch of sky. Most of the time it blends in with the rest of the pinpoints in the great wash of the Milky Way, and viewed from a city it is utterly lost in the glare of light pollution. The fact that it can be seen at all is quite impressive, as it lies 7,500 light-years away. Were the Sun pulled out to such a distance, you would need a world-class telescope like the Baade Magellan Telescope just to spot it. A look at historical records suggests that Eta Carinae was "discovered" by Edmund Halley of cometary fame, who named the star Robur Carolinum, but southern observers would have known Eta Carinae for millennia.

It seems a stretch to equate an unassuming, practically invisible speck to Thackeray's "most remarkable" object in the southern sky. Even southern observers took little note of it before 1837. That year, it became Collowgullouric War (Wife of the Crow) to the Boorong people of Australia. From total obscurity, this star suddenly rivaled the bright star Canopus (known as War the Crow) in the sky, and because it was the latter's equal, Eta Carinae was declared to be the Crow's wife. As Aboriginal astronomy researcher Duane Hamacher explained to me, the Great Eruption was the first opportunity for Western astronomers to witness in real time the creation of a story line related to a transient astronomical event.

Eta Carinae seemed to waffle for several years, growing brighter and dimmer with no apparent rhyme or reason, reaching peak brightness for a few days in 1843 and then fading again into complete obscurity by 1860. Then, visible only to telescopes, it briefly brightened again in 1887. For the next century, as Eta Carinae gradually crept back to naked-eye visibility, astronomers struggled to understand why it had unleashed so much energy so abruptly and, furthermore, why it wasn't dead. Even with new tools like radio, infrared, X-ray, and ultraviolet telescopes, Eta Carinae remained a puzzle.

— There's the Kaboom —

"So what the heck is going on inside this thing?" I asked Smith, getting right to the point.

The question might have been straightforward, but the answer is most assuredly not. For decades Eta Carinae was classified as a "luminous blue variable," and its outbursts were apparently just things that happen to such stars from time to time. Luminous blue variables are, as their name suggests, extremely bright (luminous), extremely hot (blue), and extremely badly behaved (variable).

Eta Carinae appears to us as red, however, so it could seem that the moniker is utterly inappropriate. We don't see the whole story with visible light, though. If we view Eta Carinae in infrared wavelengths, it is the brightest object in Earth's sky outside our own solar system. The light from whatever lies within heats up its cloak of billowing gas and dust, and in turn that gas and dust glow brightly in the infrared. On the other hand, when we look in the ultraviolet, we see rays of light peeking through holes in that cloak, shining like sunbeams through clouds. Those energetic wavelengths tell astronomers that a very blue, very massive monster hides beneath the mess of material that Eta Carinae has coughed up during its tempestuous life.

At first, astronomers were quick to blame a single star with a mass of well over 100 times the Sun's heft for all its outbursts. Its central fusion engine, they reasoned, must have been pumping out so much energy that it was tearing the star apart. But why that single massive star erupted with a billion years of sunshine—or a tenth of a foe—in 1843, spitting out more than 15 Suns' worth of material in the process but *not* actually exploding as a supernova, was a bit fuzzy. By all accounts, it released more energy than the explosion that created the Crab Nebula, and we see how well things turned out for that star.

Smith and his colleagues, along with plenty of other researchers, have been trying to piece together the puzzle for years, and some of the newest clues have come from a convenient cosmic do-over. Or if not that, at least a partial eyewitness account from a neighbor. When Eta Carinae became a supernova impostor, light from that event raced off in every direction. Some of the light came straight toward us, but some headed in a more "sideways" direction toward a neighboring cloud of gas and dust, which acted as a hazy cosmic mirror. The speed of light is fast, but not infinitely so, and the light that took the shortest path took less time to reach us than the light that visited the neighbor first. After more than a century, the straggler photons that took the scenic route finally arrived, giving astronomers like Smith a fuzzy replay of the nineteenth-century eruption.

"This is super cool, by the way," Smith interjected. "It's like a time machine."

And that's a very handy feature because astronomers of the 1840s lacked the vitally important tool of spectroscopy, so information about the motions and conditions of the material ejected during Eta Carinae's eruptions was seemingly lost. By analyzing several years of these light echoes, today's astronomers have concluded that there was definitely an explosive event that propelled material at speeds only ever seen in a bona fide supernova. This evidence, along with decades of observations in every wavelength regime, has helped astronomers piece together a probable story, which involved a violent merging of stars that triggered the explosion.

"Eta Carinae must have started out as a massive triple system, with two stars close together and a third farther away," Smith explained. The story that seems to work the best begins with a tight binary containing one star with 60 times the Sun's mass and another star with 40 times the Sun's mass. Sweeping around the innermost pair was a star of 50 solar masses. For a million years or so, things were fine for this trio. Unfortunately, stars eventually use up their fuel and swell past their gravitational territory, and this is where things started to go awry for Eta Carinae.

"Mass from the 60-solar-mass star began to spill over to its partner, stripping off the outer layers of the first and increasing the mass of the second. As the mass transfer between the two inner stars ended, their orbit widened until it started to interfere with the third star, which until now had been happily going about its business," Smith explained. For thousands of years, the trio was a churning, tangled mess of ejected material, mass transfer from one star to another, and chaotic orbits. It's even likely that the outer star of the trio cut in on the increasingly muddled dance, kicking what was left of the original highest-mass dance partner to the outside. From there, the two new partners whipped around each other in an ever-tightening orbit. Eventually they morphed into two cores sharing the same outer envelope, whirling faster and faster, closer and closer, until . . .

Kaboom!

Eta Carinae became everything that KIC 9832227 could not. The energy from the merging cores blasted a shock wave into space, slamming into the blobs of gas and dust that had been coughed and wheezed out of the system over the millennia. Since the cores still had plenty of fuel for fusion, they did not collapse. Instead, they created a single superstar packing at least 100 times the mass of the Sun. Meanwhile, the ejected dance partner, now whittled down to about 30 Suns of matter—half of what it started with—fell into the highly elongated orbit that is seen today. Every 5.5 years, this rejected, diminished suitor swoops around the now-coalesced cores, and in the process, it kicks up energetic blasts of X-rays that astronomers have noted for years.

It's no secret that both survivors of this mess are on borrowed time. One star has already gone through most of its life with twice its current heft. Even after having half its matter removed, it's on a fast track to destruction. At best, it might enjoy another few hundred thousand years of life. Its more massive partner has even less time. Given that we see the Eta Carinae system as it was 7,500 years ago, there is a decent chance that the monstrous star at its center is no longer there.

Unfortunately, there's no telling when we will find out how this saga ends.. It could be tomorrow, for all astronomers know. Stellar deaths have a long and storied history of sneaking up on us.

CHAPTER 10

Cloudy with a Chance of Neutrinos

Great storms announce themselves with a simple breeze.
—BISHOP OF AQUILA, *LADYHAWKE*

— Prelude —

On 23 February 1987, minus 168,000 or so years for light travel time, Sanduleak −69 202 was enjoying its last day as a self-contained star in the Large Magellanic Cloud. Well, perhaps not "enjoying." At that point, it had already eaten through its core hydrogen, converting it into helium. Then it ate through the helium, creating carbon and oxygen. Successive layers of reserve tanks within the star generated energy through the fusion of heavier and heavier elements. But like a certain caterpillar of childhood book fame, it was still hungry, and it eventually squeezed as much energy out of the atomic nuclei in its heart as it could. Its last meal—silicon, which was fused to iron—took about a day of its several-million-year lifetime to consume.

Seen from the outside, there was no indication that Sanduleak −69 202 had reached the end of its fusion rope. At the time, astronomers were generally in agreement that the final stage of a massive star's life before core collapse would be a red supergiant, a star whose outer envelope is so bloated that astronomer Robert Burnham once called it "a red-hot vacuum." Betelgeuse, located at the armpit of the great hunter Orion, is such a star. Although the star itself is just shy of 20 times as massive as the Sun, Betelgeuse could nest over 400 million Suns within itself. So tenuous is its outer atmosphere that it's impossible to say where the star ends and space begins. For decades, Betelgeuse was on the great cosmic watch list, as astronomers generally agreed that it *was* near the end of its fusion rope and, consequently, would host its grand finale any second.*

But Sanduleak −69 202? It had been cataloged in 1970 by astronomer Nicholas Sanduleak and declared a blue supergiant. Not even as intriguing as a luminous blue variable like Eta Carinae, it was filed away as something that would not likely do anything drastic any time soon. The star was slated to be observed in some detail in December 1986, not because it was particularly outstanding, but because it was one of several blue supergiants in the Large Magellanic Cloud that astronomers had been keen to study. Tragically for the astronomical community, against which weather and instrumentation conspired, researchers never quite got around to it, and anything that might have helped set Sanduleak −69 202 apart observationally from its blue supergiant siblings was soon obliterated.

— Showtime —

On 23 February 1987, comet- and asteroid-hunter Rob McNaught called it a night, making the decision to put off looking at his images of an ongoing sky survey until the morning. After all, asteroids weren't going to do anything dramatic in the next several hours, and a well-rested astronomer would be better able to spot them in the photographs.

Naturally, on that date, Sanduleak −69 202 became the first naked-eye supernova witnessed in four centuries.

As impressive as the event itself was, it was not a particularly impressive sight. It was certainly nothing like the supernova of 1006, which lit up the sky for months, but it was easily noticeable to anyone familiar with the Large Magellanic Cloud. Nearly simultaneously, observers in Chile, New Zealand, and Australia noted a blue pinpoint of light where none had been earlier, shining just at the threshold of human vision.

SN 1987A was announced to the world, after double- and triple-checking, through International Astronomical Union Circular No. 4316 on 24 February 1987. It was, at the time, cautiously described as "os-

---

* More recent work has seemingly put that show on hold for a few millennia.

tensibly a supernova" in the same location as "a blue star . . . not obviously variable during the past century." Observations of that region of the LMC just two days earlier had given no hint that anything was afoot.

It was the moment theoretical astronomers and observational astronomers alike had been waiting for. Yes, it would have been more convenient had the supernova been closer, but even at arm's length, it became too bright for some of our instruments.

But there was so much more to SN 1987A than met the eyepiece.

— The Great Escape —

Ten to the power of 57. That is, give or take, how many protons join with electrons to form neutrons during the collapse of a massive star's core, and with them come equal numbers of neutrinos to satisfy the universal accountant. It's the stellar core's last-ditch effort to relieve the incredible pressure, but the relief doesn't last. In a flash, a volume the size of the Moon shrinks to the size of a city, leaving the outer core at a loss. As the rest of the core tries to collapse inward, even more electrons are thrust into the protons, creating more neutrinos, and the collapse continues. Light, following Einstein's energy-mass equivalence, becomes particles and antiparticles, all of which are frantically interacting with each other to shed the intolerable energy. Soon the neutrino numbers have increased tenfold. Now there are ten to the power of 58.

One with 58 zeroes after it.

At this point, the center of the star has become an enormous, roiling, hot atomic nucleus. Neutrinos might see the rest of the material universe as an open window, but punching their way out of a city-sized neutron inferno is another matter entirely. Each of these newly generated neutrinos packs about the same energy as a trillionth of a crashing dinner plate, individually unimpressive but collectively beyond imagination. Together, they tear at the shrinking core with nearly 100 foes of energy, forcing great fluid blobs in the core to slosh back and forth and pushing at the collapsing walls like a terrified animal in a cage. That's a trillion years of solar energy yearning to break free from something that is quickly becoming smaller than a city and denser than a mountain per teaspoon. All of this frenzied activity occurs in the blink of an eye, the fluid blobs churning to and fro at nearly the speed of light. So much is going on that it takes months of supercomputing time to render just half a second of activity, but it's the most important half second in this star's life.

The neutrinos do ultimately succeed in punching their way out. The push from the neutrinos is, it turns out, the tipping point for a supernova, as they carry with them 100 times the energy that the light does. It seems at first glance as though the tsunami of high-energy light and the shock wave from the core's sudden halt should do the trick, just as an enormous swell might look as though it will come crashing onto the beach and sweep away your sandcastle. But very often by the time that enormous swell makes its way to the sand, its energy has been sapped by the retreating wave that came before it. Instead of the anticipated frothing spray of saltwater, the wave becomes a whisper, a swish that barely creeps up the sand.

Without neutrinos, the outward surge of matter and energy accompanying core collapse meets the same fate. A rush of energy from the core promises to blast the envelope of the star into space, while the simultaneous infall always manages to quash the energy. But when conditions are this extreme, and neutrino numbers are this great, their infinitesimally small chance of interacting is all it takes to tip the energy scales. Without that extra neutrino push, gravity wins.

It doesn't take long for conditions in the star to once again become an open window for the neutrinos. Seconds, really. As they shove the wave outward, the material becomes less dense, more transparent to these ghost particles, which barely notice a trillion kilometers of lead. Within a few heartbeats, neutrinos are free to race out of the star ahead of the slower-moving, broiling-hot shock wave. Because they are nearly—but not completely—massless, their speeds nearly match those of the photons, and they get a head start of hours to days. The difference between a photon's speed and a neutrino's speed is, at most, only about one part in a billion, so the journey of 168,000 light-years from the star formerly known as Sanduleak −69 202 was not nearly enough for the light to catch up. The neutrinos from that event arrived on Earth a few hours before the light did, spraying Earth with 10 trillion quadrillion neutrinos.

What this means is that every single person alive at the time was dosed with about a quadrillion of these energetic subatomic particles, and nobody batted an eye.

A few days later, though, there was quite a bit of eye batting.

— Ghost Hunters —

It began with an urgent fax. "Sensational news!" it read. "Supernova went off 4–7 days ago in Large Magellanic Cloud. . . . Can you see it? This is what we've been waiting 350 years for!"

The "it" that they were looking for was not the light from the event, which was apparent enough to anyone who could see the LMC. The question was whether the neutrino detection experiments had made any unambiguous detections of neutrinos from the event. Unlike the electromagnetic counterpart, the neutrino message from SN 1987A was anything but obvious. Wolfgang Pauli, the scientist who had predicted the neutrino's existence in 1930, knew this would be the case. But his assertion that it would be impossible to detect neutrinos only demonstrated that he had never witnessed someone try to win over a stray cat. When something has no interest in interacting with you, sometimes you have to get creative.

So scientists got creative. An early method of spotting neutrinos involved an exceedingly rare interaction between a neutrino and a chlorine atom. If that interaction took place, the chlorine atom would become radioactive argon, since one of the neutrons in the original chlorine nucleus would convert to a proton in the process. It was estimated that a given chlorine atom could happily survive over 100 billion billion billion years without ever being altered by a neutrino. If, however, an experiment contained 100 billion billion billion chlorine atoms, about one per year would be affected. And if you increase the number of chlorine atoms by a factor of 10 or 100 or 1,000, you now have a fighting chance to catch some newly converted argon and, from that, infer that neutrinos had come through the area.

The 1960s saw the first such experiments to detect neutrinos, not from supernovae but from the Sun, where the process of fusing hydrogen to helium creates a veritable firehose of the tiny neutral particles spewing in every direction. Scientists figured that a chamber with 100,000 gallons of cleaning fluid, equating to about 20 trillion trillion trillion (2×10^{37}) chlorine atoms, should be sufficient. But because there are so many other processes that mimic an interaction with a solar neutrino, it was better if the entire project were carried out deep underground. A kilometer and a half of intervening rock, scientists figured, would shield the experiment from the worst of the stray signals.

Thankfully, there was already an empty chamber at that depth that met their needs, and it was one that people weren't exactly competing for. South Dakota's Homestake Mine, a century-old gold and silver mine, foreshadowed the neutrino experiment that would later occupy one of its tunnels. The mine was profitable not because there were rich veins of gold, but because miners carved out prodigious amounts of rock that possessed very little gold per ton. Lots of material eventually yielded the desired reward. It's fair to say that neither the mining company owners nor any of their employees ever suspected that particle physicists would eventually descend to the deepest reaches of their mine to hunt for invisible ghost particles that eight minutes earlier had been created in the heart of the Sun.

This particular experiment, which netted about three neutrino interactions per week, had a few downsides. One was that there was no way to get specific timing information on any given neutrino interaction. All the researchers knew was that a handful of radioactive argon atoms were bubbled out of the tank every few weeks. Another complication was that the experiment could probe only one of three types of neutrinos, but these particles have the uncanny ability to switch identities midflight.

Later experiments would capitalize on an even more bizarre interaction between neutrinos and matter, but they were not tuned to one particular neutrino flavor. As an added bonus, there was no need to invest in tons of dry-cleaning fluid. Instead, researchers would see the spectral, glowing auras left when energetic neutrinos interact with subatomic particles in water, briefly punching those particles to faster-than-light speeds.

Of course, everyone knows that nothing can travel faster than the speed of light. But that's light in a vacuum. Once light hits a medium like glass or water, it effectively slows down. Light loses about 25% of its speed as it travels from a vacuum to water, but neutrinos—objects that see a light-year of lead as almost a perfect vacuum—don't even notice the change in scenery. In this new medium, it's possible for a neutrino to interact with matter and give rise to a faster-than-light charged particle. And just as an object traveling faster than the speed of sound results in a sonic boom with a cone-shaped shock front, faster-than-light travel generates a similar peculiar luminous effect.

Marie Curie noted the resulting eerie glow in 1910. As the high-energy particles emitted from her samples of radium hit the edges of their glass containers, they suddenly outpaced the light. It wasn't until nearly three decades later, though, that Russian student Pavel Cherenkov set out to investigate the cause of the blue glow in his own experiments. Now, the eponymous Cherenkov radiation has become another tool to understand what's going on behind the cosmic curtain.

The Kamiokande II experiment in Japan was set up to detect this convenient glow in the early 1980s. Following on the heels of its predecessor, Kamiokande, Kamiokande II was a Cherenkov detector a kilometer underground. The chamber was a cylinder the size of a small house containing 3,000 tons of water. Lining the inside walls were hundreds of beach-ball-sized electronic eyes called "photomultiplier tubes," each capable of detecting a single photon from the superluminal boom of Cherenkov radiation on the off chance that a neutrino decided to make its presence known.

Kamiokande II was not originally designed to catch a supernova in the act or even to spy on the fusion processes inside the Sun. Instead, it was designed to catch the faintest clue that the ever-faithful proton could spontaneously break apart and spit out its smaller constituents at faster-than-light speeds. But by 1986, it became clear that the experiment could be repurposed to detect neutrinos coming from our parent star.

Nobody suspected that it might catch neutrinos from beyond the Milky Way.

— A Simple Breeze —

Twelve.

Of the quadrillions upon quadrillions of neutrinos that washed through Kamiokande II's 16-meter-wide tank, there were 12 that made any mark. A third of the way around the world, in the Caucasus Mountains of Russia, the Baksan Neutrino Observatory picked up 5. And nestled 600 meters beneath the shore of Lake Erie in the United States, the Irvine-Michigan-Brookhaven detector netted 8. If you started handing out those neutrinos to the authors of a 1987 paper, "Observation of a Neutrino Burst in Coincidence with Supernova 1987A in the Large Magellanic Cloud," you would run out of neutrinos before you got through the P's. And that was just one of the neutrino discovery papers.

In the dim world of neutrino astronomy, those two dozen neutrinos lit up the three detectors like Christmas trees for a span of 13 seconds, assuming each tree is the size of a house with just a handful of lights. Hours later, observers in the Southern Hemisphere noted a new faint blue dot, but Earth had already received word from another messenger that a star's life had ended.

The light and neutrinos from SN 1987A arrived during astronomy's awkward teenage years. The light-gathering side of science had grown up enough that there were observatories with precision tools to gather information in every wavelength regime, and digitizing data was becoming the norm. Unfortunately, for both neutrino and multi-wavelength observations, much of the data was stored on enormous reels of magnetic tape. There was no Hubble Space Telescope, no internet, and virtually no instantaneous sharing of information between scientists, save phone calls. The desperate fax to researchers at Kamiokande II just after the publication of International Astronomical Union Circular No. 4316 was followed with a days-long process of locating the reel of tape with the relevant data, printing it out on a dot-matrix printer, poring over the findings to make sure everything was consistent, and ultimately letting the world know that an exploding star had in fact talked to the neutrino scientists first.

Since then, things have gotten quite a bit more sophisticated. If the three neutrino observatories that caught a sneak peek of SN 1987A had been able to relay in real time that they had been flooded with a burst of neutrinos, astronomers might have been able to catch the supernova's light much sooner. And while neutrino detectors might not be able to accurately pinpoint the source of the neutrinos, they could certainly get in the ballpark.

This is the philosophy behind the Supernova Early Warning System (SNEWS), which was conceived in the early 2000s. Pulling data from seven neutrino observatories, some of which are capable of detecting not just five or ten, but thousands of neutrinos from a local supernova, SNEWS will send out an alert to literally anyone who asks to be on its mailing list. The reason is clear. There are thousands of amateur astronomers around the world with telescopes powerful enough to spot a brightening new dot in our Galaxy, and every one of those astronomers would love to be the first to glimpse a star's explosive finale. Should a flood of ghost particles tear through our neighborhood and trigger at least two SNEWS observatories, SNEWS will alert the astronomical community within seconds. "Be on the lookout for a nascent supernova somewhere [gestures broadly] in that general direction," it will say, and soon after, thousands of eyes will turn to the sky.

Scientists are determined that the Milky Way's next supernova will *not* catch us by surprise.

— Trailblazer —

There is still the puzzling case of SN 1987A. While it was a victory for neutrino theorists, it was a source of astonishment for nearly everyone else in the astronomical community. Its first surprise was that it happened at all. Blue supergiants simply don't blow up. But this one obviously did, and as astronomers monitored the supernova's brightness over time, they soon realized that the story just kept getting weirder. For one thing, it seemed to take its time getting to its maximum brightness, and once there, it seemed strangely reluctant to dim. For another, it just wasn't that bright, cosmically speaking. Normal core-collapse supernovae can pump out up to 100 times as much light as SN 1987A did. In fact, if it weren't for the fact that SN 1987A had been so close, again cosmically speaking, it wouldn't have been a particularly interesting—or even noticeable—show.

Since 1987, over a dozen similarly odd supernovae have been discovered, making SN 1987A a pioneer for a new class of supernovae: Type II-pec, where "pec" means "peculiar."

Oddball.

Defying simple explanation.

The first naked-eye supernova in over three centuries, and it had to be an outlier.

Still, outliers give astronomers a chance to refine their models and question their assumptions. One overarching assumption was that Sanduleak −69 202 had been a single star during its life, an assumption made largely for simplicity's sake. Binary companions only complicate matters and were a last resort to be invoked only when all possible single-star models had been exhausted.

But as astronomy began to outgrow its awkward teenage phase, in 1990 the Hubble Space Telescope was launched, allowing astronomers an unprecedented view of the region near ground zero. What they found was—as was now the expectation for SN 1987A—a puzzle. At first there was just a glowing ring, but as the years progressed it evolved to look more like a luminescent string of pearls encircling the location where Sanduleak −69 202 had been.

There was some head-scratching and conceptual and computational gymnastics, but ultimately astronomers found a scenario that seemed to tie all the observations together. As the supernova blast wave raced through space, it slammed into rings of material that had been jettisoned from the star thousands of years before the supernova. Forget the fact that blue supergiants were not good supernova candidates. This object had clearly kicked out parts of itself along the way, and in a manner that looked as though it hadn't acted alone. In fact, Sanduleak −69 202 was beginning to look almost like Eta Carinae Lite, a lesser version of the hot mess that had joined two monstrous stars into one while expelling successive outer layers and giving southern observers a show in the 1800s.

A 2021 computer recipe for SN 1987A begins with two stars, one with 9 times the mass of the Sun and the other with about 15 solar masses. As they melt together, the resulting star tries to behave like a red supergiant but fails because its interior life has become such a mess.

The outside then presents itself as a blue supergiant—hotter and more compact than its cousin the red supergiant—while the inner merged core races toward oblivion. This model seems to re-create both the light and neutrino observations of SN 1987A, as well as the rings of pearls that have been slowly illuminated as the years progress.

So the single-star proponents were right, as were those insisting that it was a binary star. Even though it wasn't originally a single star, it eventually became one. But if that's the case, where is its leftover core? The resulting blue supergiant, after losing some of its matter in the merging process, contained about 20 times the mass of the Sun, and by all accounts it should have left behind a neutron star. A conveniently aligned pulsar would have been ideal, but even without a view of the lighthouse beacon, astronomers were fairly sure that they had the tools to detect the presence of a pulsar by its effects on the surrounding material. SN 1987A was close enough that astronomers had nearly front-row seats to the events unfolding around it, and yet there remained the stubborn lack of an obvious neutron star.

Finally, in 2019, the Atacama Large Millimeter Array, a radio observatory in Chile, reported that its scientists might have spotted the wayward neutron star. Well, not the neutron star itself. But there is an irregular loop of hot material that might or might not be the cocoon in which the neutron star is hiding. At the very least, it's in the right place, and the cocoon is hotter than the surrounding area, so perhaps a neutron star is energizing it.

Then in 2021, astronomers using an orbiting X-ray observatory reported somewhat more conclusive evidence of a pulsar. It doesn't point its beams in our direction, but it's a pulsar nonetheless. The data seem to indicate that there is the same type of pulsar wind that is gradually tearing the black widow pulsar's mate apart. Again, though, this is not the smoking gun astronomers had hoped for. If anything, it tells us that it might take a while longer than we anticipated for a pulsar to emerge from the rubble of its parent star's destruction.

And so we wait.

For the neutrinos heralding a new supernova.

For the pulsar of a past one to show itself.

For more things to shake up our universe.

CHAPTER 11

Not "The End"

"Yes, but I grow at a reasonable pace," said the Dormouse.
"Not in that ridiculous fashion."
—Lewis Carroll, *Alice's Adventures in Wonderland*

— Wrong Kind of Star —

Astronomer Nick Suntzeff's life was certainly being shaken up. Admittedly, his career revolved around the deaths of stars. This, however, was not what he had signed up for.

He sat in a corner of a small emergency room in a small hospital in the city of La Serena, Chile, desperately trying to phone the wife of the man writhing in agony in the opposite corner. Prison guards then brought in a blood-soaked man on a stretcher who, once inside, tumbled off the stretcher and immediately bolted out the double doors and into a waiting taxi. A gunfight broke out. Amid the screams and gunshots, Suntzeff succeeded in contacting Arlene and explained the dire situation to her. Her husband needed emergency surgery. No, there was no time to fly him to a larger, more modern hospital. Thankfully, one of the world's top physicians in the field had set up shop in La Serena, and he could do the operation.

"He looks just like the actor Alan Alda," the doctor said.

"He *is* Alan Alda," Suntzeff replied.

Two hours later, surgeon Nelson Zepeda came out with the good news. Alda was going to be fine. In a most unlikely twist of fate, Zepeda had operated on the man whose television portrayal of a surgeon had inspired him to become a physician.

Alda had been in Chile to interview astronomers for his science program, *Scientific American Frontiers*, when his medical disaster

· 113 ·

struck.* Suntzeff, then at the Cerro Tololo Inter-American Observatory, and his team were busy discovering distant supernovae, not so
much to learn more about the supernovae themselves, but to use them
as tools to understand what the universe was doing. As Alda and the
world had found out a few years before, what the universe was doing
made no sense, and he was hoping that conversations with the astronomers involved would help him understand.

Surprisingly, the supernovae that Suntzeff and his team had been
hunting were not the final acts of massive stars, single or otherwise.
Instead, they were just the cores of ordinary stars that had reached
their limit.

— Under Pressure —

After all is said and done, the only thing a low-mass star has to show
for its life's hard labor is a white dwarf. Not nearly as dense as its neutron star cousin, a typical white dwarf is about the size of Earth—a
few thousand kilometers in radius—but 200,000 times as massive.
Half a Sun, give or take, is squeezed into a space the size of a puny
planet. If you were somehow able to survive the 100,000-degree heat,
the 200,000 g's of gravity would get you.

Unlike neutron stars, white dwarfs announced their presence
and their peculiarities long before astronomers had any inkling that
they could or should exist. The bright dot known as Sirius is, it turns
out, two dots, a fact that remained hidden until the mid-1840s. It took
another 70 years for astronomers to obtain the faint companion's
spectrum, and in keeping with pretty much all of astronomical history,
they found it to be unlike anything they'd seen before.

It was clearly Sirius's binary partner, so it had to be the same distance away as Sirius. But what became known as Sirius B was mystify-
ingly hot. At 25,000 degrees Celsius, the smaller dot was 2.5 times the
temperature of its much more obvious sibling, Sirius A. This made no
sense because hotter stars were understood to be, as a rule, much,
much brighter. The relationship between temperature and luminosity

* For the curious, it was season 14, episode 5, "The Dark Side of the Universe,"
  which aired on 22 June 2004.

had been quite well established for most stars, and the only way for Sirius B to be *that* hot, and yet *that* faint, was for it to be *that* small.

Being small wasn't the problem. Lots of things are small. Earth. The Moon. Kittens. No, the problem was that the orbits of these two stars about their center of mass revealed that the tiny object must be as massive as the Sun.

The astronomical community was dumbfounded.

By 1931, though, the confusion seemed little more than a faint memory. "The possibility of the existence of matter in this dense state offers no difficulty," wrote astronomer E. A. Milne. "As pointed out by [Arthur] Eddington, we simply have to suppose the atoms ionized down to free electrons and bare nuclei."

"No difficulty" was quite a departure from the earlier admonitions of astronomers to white dwarfs, specifically, "Shut up. Don't talk nonsense!" Somewhere between those two sentiments arose the bizarre and unintuitive field of quantum mechanics, which revealed what subatomic particles can and cannot tolerate. Being crammed into a ball weighing a ton per teaspoon is doable as long as the electrons and atomic nuclei are properly arranged.

But it was also determined that there is only so much straw you can put on this quantum camel's back before the camel explodes.

— Core Dump —

A white dwarf is a densely packed inferno made of the last things its star was able to cook up, which, for something with the Sun's mass, would be carbon and oxygen. Like a neutron star, a white dwarf has layers, but a white dwarf's layers are made of things that seem at first glance to be familiar. Oxygen, the hardest nucleus for such a star to fuse, tends to occupy the innermost core, which is the only place where conditions extreme enough to forge oxygen exist. Meanwhile, carbon fills most of the rest of the volume. A crust of helium resides in the top couple of hundred kilometers, and occasionally some residual hydrogen might hang out above that.

Unlike a typical star, a white dwarf's size is not governed by the balancing act between the pull of gravity and the push of hot gases and photons jostling to break free. Instead, its size is dictated by the dances of the electrons, which is the same issue that the helium core experienced just before its energetic flash. This arrangement makes for some seemingly backward behavior. Add more mass, and a white dwarf paradoxically becomes smaller. Take mass off, and it grows. Add heat, though, and it doesn't change its size at all. It simply heats up.

And that's where the problems arise. When a star hits a certain stage in its life, it feels a tremendous, growing pressure in its heating helium core, a condition that sparks a brief episode of runaway fusion. Releasing the equivalent of millions of years of solar energy in a minute, the helium flash foreshadows what could happen . . . "if," the star says menacingly, "certain conditions are met."

A white dwarf is a powder keg waiting for that spark. Alone, it will be fine (for the extreme foreseeable future). But a partner has the potential to set it off. The easiest way to do so is for the partner star to age, swell past its Roche lobe, and drop some of its hydrogen envelope through the narrow gravitational doorway between the two objects. The hydrogen races toward its dead partner's heart, heating as it goes, and collects on the surface of the white dwarf. Feeling additional pressure, the white dwarf becomes smaller and hotter.

This process can go on for quite some time—tens of thousands of years, even—but it can't go on indefinitely. Once the surface hits a suitably scorching temperature of 20 million degrees Celsius or so, the hydrogen does the only thing it can. It fuses into helium, releasing a flood of energy that makes the densely packed hydrogen fuse even more efficiently, releasing a bigger flood of energy that makes the hydrogen fuse even more efficiently, and so on. Within seconds, this runaway thermonuclear process has turned the entire atmosphere into an enormous hydrogen bomb. The temperature skyrockets to over 300 million degrees, and the surface that the white dwarf has been so patiently amassing is blasted out into space at breakneck speeds exceeding 3 million kilometers per hour. At this rate, it would take just five seconds to get from London to New York, but the flight would be uncomfortably hot.

Here on Earth, we see an obscure dot brighten over the course of a few hours, and then slowly fade away over the next several days as the superhot, superfast shell cools and dissipates.

A nova (not a supernova) is born.

At the end of this surface explosion, this scrappy little dead core of a low-mass star will have spat out about a millionth the energy of a typical supernova, or a micro-foe, along with a couple of teeth. Not bad for something that doesn't even leave a scar. Depending on the dynamics of the binary system, the two characters might reprise the play some day in the not-so-near future.

While 10,000 years is a long time to wait for the next outburst in a system like this, astronomers have options if they want to observe a nova. Given that (1) a galaxy like ours contains hundreds of billions of stars; (2) low-mass stars that end their lives as white dwarfs are extremely common; and (3) binary systems are the norm, novae are popping off all over the universe. Indeed, it's estimated that somewhere in a galaxy like ours there is about one nova per week, most of which go unnoticed because the dark dust of the Emu and other sooty clouds obscures them from our sight.

— The Last Straw —

Deep in the heart of a white dwarf, the universe walks a fine line between stability and destruction, a fine line occupied by the supporting cast of electrons. In the early twentieth century, these minuscule electrons weighed heavily on the minds of astronomers and physicists, since so much in the physical world seemed dependent on their whims.

In 1930, Subrahmanyan Chandrasekhar embarked on his professional scientific career by literally embarking on a voyage from his home in India to England. During the journey, he refined the calculations of his soon-to-be graduate advisor, Ralph Fowler. Fowler had been determining the internal conditions of white dwarf stars, which are incredibly dense. The pressure to keep them from succumbing to gravity's crush comes from the fact that no two electrons can have exactly the same energetic qualities. It's an esoteric topic, but in his calculations Fowler failed to account for an even more esoteric topic: relativity, or the bizarre behavior of things that move incredibly fast or are subjected to extremely strong gravity. Within a white dwarf, electrons are such things, and rather than simply apply the usual laws of physics to the situation, Chandrasekhar realized that relativistic effects needed to be considered.

While most passengers basked in the light streaming off the surface of our nearest star, Chandrasekhar concerned himself with the fate of its deepest interior, a task that would earn him a Nobel Prize half a century later. His calculations revealed that a white dwarf's electrons could take only so much. As the mass of the white dwarf increases, the electrons in its heart move at a decent percentage of the speed of light. When the pressure becomes too high and the electrons become relativistic, the electrons do the only thing that's left for them to do: join with the protons to make neutrons.

Chandrasekhar discovered in his calculations that, given a certain fraction of electrons in the inconceivably dense interior, the maximum mass for a white dwarf is 1.44 times the mass of the Sun, a figure now known as the Chandrasekhar limit.* Beyond this, the electrons flee into the atomic nuclei. Once that happens, it's curtains for the white dwarf. It collapses into a neutron star, spitting out a flood of neutrinos and gamma rays in the process. The outermost layers of the white dwarf might also get blasted off.

This scenario is most assuredly *not* on the poster in every Astronomy 101 classroom. Making a neutron star is taught to be an activity strictly reserved for massive stars.

The problem with the Chandrasekhar limit is not that it's incorrect. It's that a typical white dwarf winds up tearing itself apart before it can even get to that point. Just before the electrons feel compelled to dive into the protons, conditions finally become perfect for fusing carbon into heavier and heavier elements. This kicks off not because the temperature is so great—although it is plenty hot in there—but because the pressure is so high.

Once that process begins, it's also curtains for the white dwarf, but with a different result. The center of the dead star becomes a fusion bomb that can't punch its way out of the dense ball that contains it. All

---

* Suntzeff reminded me that there was another astronomer working toward the same goal around the same time: Edmund Stoner. The mind reels at the thought of undergraduate students learning about the Stoner limit.

that energy ignites more fusion bombs, which ignite more fusion bombs. And yet again, the extreme material in the white dwarf won't budge until the runaway fusion process literally blasts the white dwarf into oblivion.

There is no neutron star left behind, or anything else for that matter. Racing out to meet the rest of the universe is a Sun's mass of new fusion products—radioactive nickel, for one—cooked up in the last milliseconds of the white dwarf's existence. After a few days, the radioactive nickel nuclei shake themselves off, flick a spare positron from one of the protons, and become radioactive cobalt. After a few months, the cobalt nuclei do the same thing to become the more stable and more common iron. All those antimatter positrons meet up with electrons and create gamma rays, adding even more energy to the expanding gases, which glow for many months before finally tapering off. When the show is over, energy equivalent to the Sun's 10-billion-year lifetime will have been unleashed. A foe of energy, emerging from something smaller than our planet.

— Typecasting —

One of the most striking features of this type of supernova explosion is that because it starts out as a colossally dense orb of carbon and oxygen, its spectral fingerprint shows no evidence of hydrogen. When a massive star blows up, though, the outer layers are typically still relatively undisturbed hydrogen, which doesn't know what all the fuss is inside. The spectra of those sorts of supernovae reveal the skirmishes between light and hydrogen.

As early as 1941, Minkowski had noticed that the spectra of some supernovae displayed wavelengths associated with hydrogen while others didn't, but it wasn't clear then why that should be the case, so he did what every good observer does. He classified them as Type I (no signs of hydrogen) and Type II (signs of hydrogen). SN 1987A would be one of the latter, albeit with the "pec" suffix. Beyond that, among supernovae that didn't show any hint of hydrogen, there were variations in both the details of the spectra and how the supernovae dimmed over time.

As for Fritz "Supernova" Zwicky, the pioneering supernova hunter, it turns out that his first 35 supernova discoveries were considered

Type I. Although there are plenty of Type I supernovae that do involve core collapse, it's possible—but not likely—that Zwicky didn't see one of these play out until his 36th catch. What is likely is that many of the supernovae he caught were white dwarfs that had finally caved in to pressure from their binary partners, a scenario he never envisioned. Nobody knows for certain the breakdown of what Zwicky saw. Astronomers wouldn't start peeling apart different classifications of Type I supernovae until the 1980s based on the exact pattern of brightening and dimming. It was then that Type Ia supernovae, the sort created when a white dwarf turns into a fiery fusion bomb, became a class by themselves.

— Not So Fast —

All this sounds well and good, but it leaves plenty of open questions. First, if we're going to blame a binary companion for the explosive demise of a white dwarf, wouldn't it help to actually have evidence of a binary companion?

In 2012 astronomers finally got confirmation that, at the very least, the progenitors of *some* Type Ia supernovae had a surviving partner. The evidence came from SN 2012cg, an event in a galaxy 50 million light-years away. At this distance, it's impossible to make out the partner visually, but there is a chance of seeing what the explosion does to the partner. In this case, having a close friend explode as a stellar-mass fusion bomb heats up one side of the companion just as sitting next to the fireplace heats up one side of a lazy dog. This heating makes the surviving partner glow with shorter, more energetic wavelengths. Through months-long, exquisitely detailed analysis of the light from this event by teams of astronomers using several observatories dotting the globe, an ultraviolet observatory *orbiting* the globe, and computer modeling of the scenario, astronomers could say with some measure of confidence that, yes, we might have seen the warm side of a dog.

It's possible, though, that a Type Ia supernova isn't always just a stellar vampire drawing material from a large dying star. After all, so much has to go right (or wrong, if you want to look at it that way) for this to happen. The most obvious issue is that somehow the white dwarf has to pull matter off its partner slowly and gently enough that there aren't any nova outbursts, events that effectively reset the white dwarf's mass. It's like trying to save up for that dream vacation when an unexpected bill empties the piggy bank. The second problem is that the matter being pulled off is largely hydrogen. If the white dwarf explodes while bathing in a sea of hydrogen, that element is going to leave behind clues that it was there, and then our supernova will be a Type II event.

Consequently, astronomers have modeled other channels to create these events. If the universe can somehow persuade two white dwarfs to merge *and* if their combined mass is sufficiently close to the Chandrasekhar limit, then there will be spectacular mutually assured destruction. What's more, this mechanism all but guarantees that all Type Ia supernovae will have approximately the same energy output, unlike core-collapse supernovae, which arise from stars with an enormous variety of masses. In a universe of messy interactions, extreme conditions, and epic explosions, reliability is a very useful feature indeed.

— Supernova Slump —

Reliability allows astronomers to use Leavitt's law to gauge cosmic distances, divining the energy output from timing a Cepheid variable star's ups and downs. Even as early as the 1940s, Baade and others had seen hints that the exact progressions of brightening and dimming of some supernovae had very similar properties. In the late 1960s, after dozens more supernovae had been analyzed, astronomer Charles Kowal declared, "It is obvious, therefore, that these supernovae could be exceedingly useful indicators of distance. It should be possible to obtain . . . accuracies of 5% to 10% in the distances. The main problem now is one of calibration."

For astronomers, the implications of this paper were enormous. A supernova is as bright as the entirety of its host galaxy, which can harbor hundreds of billions of stars. Essentially, if you can see the galaxy, you can see a supernova in that galaxy, assuming one pops off. If the supernova is of the right parentage, something that can be determined by looking at its spectral fingerprint and light curve, then you stand a chance of knowing exactly how much energy is being emitted. And, as with so many of astronomy's standard candles, knowing an object's energy output and measuring how bright it appears from Earth leads to an object's distance.

Since the 1920s, astronomers had known that the observable universe is expanding in all directions. It is as though galaxies are riding along on a stretching elastic medium, each galaxy seemingly rushing away from the others at breakneck speeds. Closer galaxies recede more slowly (that is, they have smaller redshifts) than more distant ones, just as the separation between two dots on an inflating balloon increases more if the dots start out farther apart. Decades of observing the redshifts of galaxies and applying Leavitt's law, as well as using a variety of less reliable distance indicators for the truly distant galaxies, had supported this universal expansion.

But what was sorely lacking were precise, independent distance measurements, particularly for the most distant galaxies. Was the expansion rate of the universe so great that if you wound time backward, it was a mere 10 billion years old? Or was the expansion rate small enough that the universe could have taken as much as 20 billion years to get to its present state? Moreover, was the amount of gravitating matter in the universe enough to halt the expansion and bring the universe back together at some point in the extremely distant future? Or had that initial bump in the universe known as the big bang been so energetic that the universe was destined to keep flying away from itself forever?

Complicating matters were the pesky observations and models of stellar evolution that seemed to insist that the oldest stars in the Milky Way Galaxy were older than the actual universe. Clearly the children can't be older than their parents, but no scientists could see where they'd miscalculated. Between different subfields of astronomy there was a marked tension. Astronomers fell into philosophical camps, arguing that the other side had failed to account for X, Y, or Z.

Someone was definitely wrong about something—possibly a whole lot of somethings—and a bright, reliable standard promised some kind of resolution. By the early 1970s, astronomers were discovering only about two supernovae each month, a rate that was convenient only for those who were naming the events. Supernova designations reveal the year (e.g., 1987) and the order of discovery (e.g., A for the first one of the year), so with fewer than 27 annual supernova observations, at least astronomers didn't run out of letters.

But not all those supernovae were Type I, the sort that might possibly be useful distance indicators. Indeed, fewer than half of them were ever assigned a type, so inconclusive were the messages in their rainbows and changes in brightness. Worse, few had ever been spotted at what could be considered cosmological distances beyond hundreds of millions of light-years.

This was hardly progress.

Interest in hunting exploding stars waned as other shiny topics drew astronomers' attention. It was a thankless process after all. Photographing distant galaxies with the hope that one would show a new dot seemed to be a poor use of not just telescope time, but human time. It took considerable effort to expose photographic plates, develop plates, and then visually compare plates. As much as the field of astronomy had grown up since the days of the Harvard computers, some things were hopelessly stuck in the past.

Then the digital revolution hit. Images could be taken using the new charge-coupled devices (CCDs), and computers could do the heavy lifting of scanning for new dots. The Berkeley Automated Supernova Search was kicked off in 1981 with a state-of-the-art 500 pixels × 312 pixels sensor, 512 kilobytes of memory (more than enough, the proposal assured the reader, as each image would only be 320 kilobytes in size), and two—count 'em—*two* 10-megabyte disks. To put this in perspective, today a single phone camera image of a mouthwatering pasta dish would practically fill the entirety of the automated search's data storage.

The method for the search was incredibly basic. Take a galaxy image. Read the image into the capacious 512-kilobyte memory. Move the telescope to another galaxy. Meanwhile, the custom software analyzed the first image, looking to find any hint of a supernova. If it found nothing, then the telescope simply moved to yet another galaxy, and the process was repeated. Rather than poring over everything visually, astronomers could let the computers and digital detectors do the work, accomplishing in minutes what used to eat up hours or days.

It seemed like a slam dunk, a way to catch exploding stars and better understand them and possibly even better understand the entire universe. Paradoxically, the astronomical community regarded the project with apparent indifference. The shiny object of the day was the Big Picture. Despite the advertised promise of supernovae, they didn't seem to be all that helpful in answering the big questions.

— All for Naught —

"This is what was drilled into my mind," Nick Suntzeff said. "My mentor Allan Sandage would say, 'There are only two numbers in astronomy: H naught and q naught. And that's what you should be working on, Nick.'"*

H naught, written H_0, is the Hubble constant, and it's the relationship between a galaxy's distance and its redshift. In 1929 Edwin Hubble found that the redshifts of galaxies become greater as we observe more distant galaxies. One of the first hints that our universe is expanding and dragging the galaxies along for the ride, this clear correlation can be seen in a graph of the recessional velocity (measured in kilometers per second) and distance (measured in megaparsecs, where 1 megaparsec = 3.26 million light-years). The slope of that graph became known as Hubble's constant, and it's reported in kilometers per second per megaparsec.

This is anything but an intuitive unit, and its value is anything but settled. Hubble originally estimated the slope of the graph to be a whopping 500 kilometers per second per megaparsec, indicating that two galaxies that are one megaparsec away from each other would be separating at 500 kilometers per second; a pair that began two megaparsecs apart would separate at 1,000 kilometers per second. Since Hubble's time, better observations have gradually pushed that number down to about 70, give or take a few kilometers per second per megaparsec.

* Indeed, in 1970 Sandage wrote an entire article about the importance of these two numbers to cosmology, just in case Suntzeff refused.

It's the "give or take" that has caused considerable consternation in the astronomical world. Astronomers have not been able to definitively zero in on the Hubble constant's value, and by extension we are missing something fundamental about the growth history of the universe. You see, the Hubble constant isn't simply the slope of some curve on a graph. If you know the Hubble constant, you can determine the age of the universe by essentially running everything backward until you hit the x-axis. Unfortunately, the Hubble constant is not, as its name would suggest, constant. The value of 70 give-or-take kilometers per second per megaparsec applies only to the here and now. In the era of the first galaxies, the figure was much higher. Because of its relationship to the overall size of the universe, the Hubble constant (more accurately, the Hubble parameter) has been decreasing throughout the history of the universe, as astronomers expected. Precisely how it's been decreasing, though, is hard to nail down, and that's where q naught comes in.

Q naught, indicated by q_0, is the deceleration parameter, an equally important value. It tells astronomers how much the universe's expansion is changing. A positive value, and the expansion rate is slowing. A negative value would mean that the expansion rate is increasing. While fights about H_0 were being fought by astronomers worldwide, the value for q_0, everyone agreed, was definitely positive. Absolutely, positively, undeniably positive. It couldn't be otherwise.

To understand why, just imagine throwing a ball straight up. As it gets higher, the ball gradually slows. If you clock its speed along the path, you find a very specific relationship between the height of the ball and its speed. Even if you threw the ball so quickly that it launched off the face of Earth, it would still lose speed as Earth's gravity sapped its energy. Galaxies should be doing the same thing. Sure, the space between them might be both incomprehensibly vast and rapidly growing, and sure, the universe does have a nasty habit of throwing weird and unexpected things our way, but this one seemed nonnegotiable. The enormous amount of stuff in the universe—billions of trillions of Suns' worth of stuff—simply had to be putting the gravitational brakes on itself. But the only way to know for sure how strong those brakes were was to clock the motions of different galaxies and get precise distances.

Plenty of people had tried throughout the decades, but the uncertainties were always too great to put much stock in any of the values. And then SN 1987A happened, and a series of lightbulbs went off over the heads of astronomers like Suntzeff.

— Acceleration —

The blast of SN 1987A launched a thousand new research projects and about as many careers. One of those belonged to astronomer Mario Hamuy, whose job at the Cerro Tololo Inter-American Observatory in Chile began just two days after the Large Magellanic Cloud sported a bright new dot. Eager to get in on the supernova wave, he teamed up with Nick Suntzeff, Mark Phillips, and José Maza to launch a new supernova survey.

"We were just a bunch of astronomers hot on the track of supernovae," Suntzeff recalled, and they had one goal: figuring out the elusive deceleration parameter, q_0.

For this task, Type Ia supernovae came to the rescue. Astronomers still weren't sure whether these resulted from two merging white dwarfs or a white dwarf siphoning material off a companion, but in either case, they seemed to be the right tools for the job. Unfortunately, they turned out not to be the uniform standard candles originally advertised. That would have been too convenient, and the universe seems to think that astronomers appreciate a good challenge.

Over the next few years, the members of the Calán/Tololo Supernova Survey discovered dozens of supernovae, more than 30 of which were Type Ia. This was a big enough sample to reveal the more nuanced behavior of these objects.

"It's a very long, overly complicated explanation, but there was no 'aha!' moment in any of this," Suntzeff said. He and his group gradually realized that by closely monitoring the exact pattern of brightening and dimming, they could precisely determine a Type Ia supernova's peak energy. From there, Suntzeff explained, "it was just unraveling the observational project and then applying that to farther and farther objects."

Once they had that puzzle piece sorted out, the rest fell into place. Now they could both clock the speed of the ball and measure its height.

By the early 1990s, Suntzeff had built a CCD camera for the telescope and fine-tuned the process, determined to apply the group's work to objects even deeper into the universe. The answers they sought were still beyond their grasp, but the team knew that with just a few more years of observing supernovae, they would finally do something that nobody else had managed to do.

But they needed a name. The team pondered the possibilities. "Well, what do we do?" Suntzeff recalled of the conversation. "We look for high redshift supernovae. So let's call ourselves that." Z is the astronomical designation for redshift, and so they became the High-Z Supernova team. The newly designated High-Z Supernova team took on new members, and Australian astronomer Brian Schmidt became its leader. It also took on new rivals, many of whom would use the same equipment that Suntzeff had built for his team's project.

"I would be up there with the support scientists finding supernovae for all the competing projects," Suntzeff mused. "One night would be mine. The next night, theirs. I'd be up there the whole time, helping my competitors try to disprove or beat us. It was competition, but it was a fun competition."

Finally, in 1998, the team was ready to share the results with the world. The title of the paper—"The High-Z Supernova Search: Measuring Cosmic Deceleration and Global Curvature of the Universe Using Type Ia Supernovae"—didn't hint at anything unusual. The implications of their findings, though, were mind-boggling.

Suntzeff chose his words carefully. "What we found was that distant supernovae are fainter than they should be. They're farther away, which means that over cosmic time, something has pushed them farther away than we thought they were going to be."

After much calibrating and considering all the possible uncertainties, the High-Z Supernova team reported a value of q_0 that was negative. The expansion of the universe wasn't *de*celerating, at least not over the past few billion years. It was *ac*celerating. This was as puzzling as finding out that the ball you threw was getting faster and faster as it went up. And that meant that something—they were careful not to say what—had to be pushing the universe apart. Moreover, it wasn't just the High-Z Supernova team that got this inexplicable result. A competing team, led by Saul Perlmutter, reported the same behavior.

It was these cosmic shenanigans that had intrigued Alan Alda enough to prompt his visits to astronomers in Chile and other locales for *Scientific American Frontiers*. Suntzeff's team even invited Alda to take over an observation, confident that he'd find at least one supernova.

Then disaster struck.

"He never did find a supernova," Suntzeff recalled somewhat regretfully, although Alda did get a consolation prize. "We inducted him into our supernova team, and we printed up something official-looking for him."

Alda's fascination with a universe that is doing the complete opposite of what astronomers set out to measure mirrored that of the entire scientific community. In 2011, the Nobel Prize Committee recognized the revolutionary discovery by awarding the Nobel Prize in Physics to the two competing teams. Unfortunately for Suntzeff and many others in those groups, the rules of the Nobel Prize dictate that the maximum number of recipients for any given award is three, a rule that seems starkly out of touch with the modern era of vast scientific collaborations that would fail without the creativity and expertise of many contributors.

Meanwhile, the scrappy little remnants of low-mass stars could stand proud, having been the focus of two Nobel Prizes: one for determining their breaking point and the other for exploiting that breaking point. The second prize even sports a poetic twist. Given that at least some fraction of Type Ia supernovae are the result of colliding white dwarfs, a small-scale coming together revealed a colossal pushing apart. But this wasn't the first time that small-scale mergers let astronomers in on an enormous universal secret.

CHAPTER 12

Collision Course

> Wave after wave, each mightier than the last,
> Till last, a ninth one, gathering half the deep
> And full of voices, slowly rose and plunged
> Roaring, and all the wave was in a flame.
>
> —ALFRED, LORD TENNYSON, *THE COMING OF ARTHUR*

— A Distant Storm —

Admittedly, Rottnest Island was not the main reason I had come to the Perth area. My official destination was the Gravity Discovery Centre in Gingin, an hour north. Unfortunately, I was going to have to miss out on visiting a radio telescope array that looks more like a swarm of mechanized spiders than an astronomical observatory. The 4,096-unit spider brigade, the official name of which is the Murchison Widefield Array, is a few hours' drive beyond Geraldton, which is an hour's flight from Perth, which is already at the ends of the Earth.

In a place as sparse as Western Australia, it's a wonder that any two things ever come together. In a universe that's even bigger and more sparsely populated—and getting bigger and more sparsely populated by the second—it seems an impossibility.

But on a sunny day off, smiling quokkas ("The world's happiest animals!" assured the tourism videos) beckoned. Twenty kilometers off the coast of Western Australia, Rottnest Island gets the full brunt of thrashing Indian Ocean waves. To get there requires a 45-minute fast ferry ride that slides up and down the swells. On this particular autumn day, the swells were rather pronounced because of a disturbance off the coast of Africa a few days earlier.

Yes, Africa. There is nothing but the vast Indian Ocean separating that mighty continent from Perth, fully 8,000 kilometers to its east.

Nasty weather near Nairobi can bring seasickness to Australian ferry riders within a week, energy from a distant storm driving the swells an ocean away.

The ferry hummed rhythmically over the swells, and as I gazed out at the undulating sea, with small ripples superimposed on the rolling hills of water, my thoughts naturally turned to binary stellar corpses, decaying orbits, and the invisible tapestry in which we live. As one's mind often does . . .

— The Invisible Energy Thief —

In 1916 Albert Einstein published his general theory of relativity. In it, he revealed our universe to be an interwoven fabric of matter, space, and time. It took less than a year for him to further calculate that when matter moves, the rest of the fabric—space and time—ripples in response. Those resulting ripples are known as "gravitational waves," and like any wave, they transfer energy from one place to another. Rolling waves near Rottnest can signal the energy dumped by a storm system near Africa, just as a buzzing sound lets you know a fly is frantically beating its wings nearby.

Einstein himself had little optimism that any consequence of gravitational radiation would ever be observable. Indeed, he occasionally doubted that such a thing even existed, going so far as to author an unpublished paper entitled "Do Gravitational Waves Exist?" At that time, two decades after he first formalized his general theory of relativity, he had become convinced that the answer was no, and he said as much. Unsurprisingly, attempting to describe how the imperceptible fabric of spacetime is affected by accelerating masses requires novel mathematical tools whose results are frequently tough to interpret physically. His mathematics were correct. His interpretation, however, needed some help.

Once it was largely accepted that, yes, gravitational waves do exist, the question remained whether they—or anything resulting from them—could ever be measured. Gravitational radiation, unlike electromagnetic radiation, has no choice but to be subtle. Light waves arise from the interplay between electricity and magnetism, forces that are 100 trillion trillion trillion times stronger than gravity, making them much easier to observe. Those waves bounce around the universe, potentially agitating any charged particles they encounter, including in our retinas, in radio telescopes, or in orbiting X-ray observatories. As it turns out, charged particles, like two young siblings, really enjoy bothering each other, even from incredible distances.

But conversations between masses are much more subdued, their voices carrying as fast and as far as light waves, but at a much, much, much lower volume. This "volume" is not a literal loudness. These are not sound waves. Instead, interacting masses create minute compressions and elongations in the tapestry fibers, a tapestry that is stubbornly resistant to carrying any messages the masses want to send.

In this picture, spacetime is a *thing*. A substance. And it can be manipulated like one. Every substance is theoretically compressible, assuming you can apply enough pressure. Steel, for instance, is about 20 times harder to compress than wood, which is about 100 times harder to compress than rubber. The strongest, least compressible material humans have ever created is something called carbyne, which puts up an enormous fight to keep its dimensions. This substance is about 18 times more resistant to compression and elongation than steel, a figure that seems impressive until you consider spacetime.

We live on a planet that seems to glide effortlessly through something that is about 10,000,000,000,000,000,000 (10^{19}) times stiffer than carbyne. All the analogies in the world cannot begin to convey the intransigence of the cosmic scaffolding. It requires the most monumental pressure to deform spacetime, but this deformation has one very obvious consequence in our lives: gravity.

It was the well-understood behavior of masses subject to gravity that allowed Einstein in 1915 to formulate the equations that encompassed the properties of spacetime. For those who don't speak mathematics, Charles Misner, Kip Thorne, and John Wheeler neatly summarized the situation in their 1973 book, *Gravitation*: "Matter tells space-time how to curve; space-time tells matter how to move."

We had seen plenty of moving matter through the centuries. Now, thanks to relativity, we could understand the behind-the-scenes curvature responsible for those motions.

The most common way to visualize the concept of a curvable spacetime is to think of a trampoline flexing and warping with each bounce. Gleeful children never have to worry about hitting the ground below because their masses are sufficiently small, their jumps sufficiently tame, and the trampoline sufficiently taut. A child who doesn't want to make a steep dip in the trampoline need only lie down. Spreading out her mass over a greater area exerts less pressure on the fabric. The steepness and depth of the dips depends on how much mass there is and how concentrated it is. And when a child lying down on a trampoline exerts less pressure on the fabric than the same child standing upright, the curvature is even lower.

Earth, in this analogy, is a child lying down. She creates a gentle slope to the fabric of the trampoline, such that if you were to put a marble near the child, it would roll casually down and rest next to her. Because its mass and size are much greater than Earth's, the Sun's imprint would be much wider but also deeper. Concentrate the Sun's mass into a tiny package like a white dwarf, and the dent becomes narrower and steeper. Further squish it into the city-sized package of a neutron star, and the dent becomes so narrow and so steep that it's almost vertical, the result of an elephant on a pogo stick. It's not hard to imagine the ever-faster circles that a marble rolled near such a pit would trace out as it spiraled toward the bottom.

In this formulation of gravity, orbits are nothing more than the natural motion along a curved surface, no force required. There is, however, a payment to be exacted. If, for instance, two objects whirl around and around and around their balance point, they are perpetually creating undulations in the universal tapestry. Those undulations sap energy from the pair and then send off that energy at the speed of light to the farthest reaches of the tapestry in the form of gravitational waves.

Puny Earth with its year-long orbit around the Sun creates shallow ripples with such long wavelengths that only one of them washes over our cosmic neighbors each year, each wave a light-year from peak to peak. The combination of masses, separation, and orbital period means that the Earth-Sun system is losing a measly 200 watts annually through gravitational radiation. If this were the only process at play, this energy loss would gradually force Earth's orbit to shrink by about a billionth of a millimeter per year, hardly a cause for concern given that we are 150 trillion millimeters from the Sun. But our solar system is a dynamic mess with far more significant factors at play, the sum of which is actually causing Earth's orbit to grow.

Heftier and faster, the pair of stars that makes up the not-soon-to-be-red-nova KIC 9832227 creates steeper ripples with a wavelength of 11 light-hours, about the diameter of Pluto's orbit. Their greater masses and smaller separation mean that more energy is being carried away by gravitational waves. But like the Earth-Sun system, the stars of KIC 9832227 have 99 energy-altering problems, and gravitational radiation isn't one of them.

To really notice the effects of gravitational radiation, what astronomers need is a clean binary system, preferably living in very close quarters and preferably with intensely concentrated masses so that other effects are minimized.

Thankfully, the universal warehouse has something for everyone.

— Hulse- and Taylor-Made —

In the early 1970s, pulsars were the name of the astronomical game. Radio telescopes like the Dish and Arecibo were racking up pulsar discoveries left and right, but not as mere collector's items. Pulsars were the best chance of seeing extreme physics play out, a stage upon which the usual rules of Newton's physics didn't apply. "To lovers of general relativity theory like me," physicist Kip Thorne gushed, "this is a very exciting state of affairs, because relativistic modifications of Newtonian theory should be important in all neutron stars!"

So what does the extreme curvature of spacetime around neutron stars do? First, there is the potential for measurable gravitational radiation. Perhaps astronomers can't directly measure the gravitational waves themselves, but they can divine how much energy a pair of neutron stars loses over time if at least one of the pair presents itself as a pulsar. Second, if astronomers are extremely lucky, they might get a chance to see an eclipsing binary pulsar, one pulsar passing in front of the other as they orbit. Because space and time are intertwined in general relativity, the stretching of spacetime by masses means that time ticks more slowly closer to a gravitating object. If one pulsar passes in front of the other, it appears to Earthlings that someone lengthened the background pulsar's beat. And finally, if astronomers hit the line-of-sight jackpot, they can even measure how the pulse beam itself follows the deep pit in spacetime caused by its foreground partner.

But let's not get too greedy. Before any of this could be done, astronomers needed to find an actual pulsar in a binary system.

Of course, it didn't take all that long. Binary stars are the norm, and while massive stars aren't the most abundant stars in the universe, there are still plenty in a galaxy of several hundred billion stars.

In 1974, less than seven years after the discovery of the first pulsar, Princeton astronomers Russell Hulse and Joseph Taylor announced an exciting new find. It wasn't a binary pulsar, but it was a pulsar with a fairly massive unseen partner. The two (objects, not scientists) danced about each other in mere hours, a dance that caused the pulsar's heartbeat first to appear faster, then to appear slower. It was the Doppler shift, but instead of the "eeeeeeeeeyoooooooooo" of sound waves or the blueshift-redshift of light waves, astronomers measured the pulse-to-pulse timing change. Instead of picking up a reliable clock ticking every 59 milliseconds, Arecibo detected slightly more frequent pulses when the pulsar moved toward us and slightly less frequent pulses as it receded.

The universe had given scientists the perfect laboratory to test one of the most audacious claims in science. If Einstein was right, the orbiting neutron stars should radiate away energy, not as light but as gravitational waves. The very fabric of the universe would be rippling, and all researchers needed to do was watch.

And wait.

And wait some more.

There is, it turns out, a considerable amount of waiting involved in trying to measure any changes in the orbit of a pair of neutron stars that are separated by, on average, 2 million kilometers. The neutron stars are engaged in a complicated and elongated dance, whipping quickly around each other when they are near, and more casually sweeping across the dance floor when they're far. Over the course of seven years, astronomers made thousands of observations, looking for evidence that gravitational waves were siphoning over 7 trillion trillion watts of power from this system.

Along the way, they made a number of other discoveries, including that the mass of each neutron star in the system is 1.4 times that of the Sun. This in and of itself was a pretty spectacular find, but authors Joseph Taylor and Joel Weisberg seemed almost blasé about it. "It is interesting to note, in passing, that PSR 1913 + 16 is the only radio frequency pulsar whose mass has been measured."

This apparent indifference can be excused, though. What they then went on to show is that the time between each close encounter between the neutron stars—the point in the dance where they whip quickly around each other—is gradually decreasing. The dance partners are moving ever closer together, each pass taking infinitesimally less time than the previous one. With one of the most stunning graphs in the history of astronomy, Taylor and Weisberg's 1982 paper shows how their measurements are precisely what you would expect if the energy thief were none other than gravitational radiation.

Once again, a Nobel Prize would be awarded to scientists who used stellar corpses as natural laboratories. This one went to Hulse and Taylor in 1993. By then, astronomers had confirmed year after year that the pair of neutron stars was, in fact, behaving precisely as Einstein's theory predicted. (Scientists are still confirming this behavior.) "The good agreement between the observed value and the theoretically calculated value of the orbital path can be seen as an indirect proof of the existence of gravitational waves," read the press release for the 1993 Nobel Prize.

Astronomers will have to wait much longer to see the inevitable finale of this shrinking dance. It will take 300 million years for gravitational waves to carry off all the orbital energy of the two neutron stars, at which time they will collide.

At that point, the Nobel Prize press release promises, things could get very interesting. "Perhaps the violent perturbations of matter that take place when the two astronomical bodies in a binary star (or a binary pulsar) approach each other so closely that they fall into each other may give rise to gravitational waves that could be observed here."

Perhaps.

But that's not all that will be created.

CHAPTER 13

Fallen Stars

And she tried to fancy what the flame of a candle is like
after the candle is blown out.

—Lewis Carroll, *Alice's Adventures in Wonderland*

— Plotting a Star's Demise —

On my side of the planet, it was nearing nightfall, a steady rain ushering in the darkness a bit early. For Nikhil Sarin, though, it was a sunny Stockholm morning at NORDITA (Nordic Institute for Theoretical Physics) when we began our video call. If anyone could explain the fate of a binary neutron star system, I figured that he could, since he had recently defended his award-winning dissertation on the topic. We fought through some technical difficulties so that Sarin could share his computer display with me, and after a few keystrokes, he excitedly presented a screen filled with—I squinted—graphs?

It wasn't quite the swirling simulation of merging neutron stars that I had been hoping for, but I was game to understand the data nonetheless.

"These plots show you how the luminosity in hard X-rays and also gamma rays changes over time," he explained. The objects emitting the most energetic types of light are over a billion light-years away. That anything about them could be observed was astonishing.

"So take a look here." He waved the cursor over a place where the intensity of the X-ray emission had initially dropped a bit on the graph, but then leveled off. "The X-ray plateau lasted, in this case, at least a million seconds. We believe that this is a neutron star that survived a binary collision and was actually ultimately stable."

"What if it doesn't survive? What happens to it?" I asked.

Sarin made cursor circles around another graph. "Look at this one in the middle. There are some X-rays lasting up to 100 seconds, and then all of the sudden, they just seem to switch off." He seemed particularly enthusiastic about this event, but it was impossible to tell why.

"What does that mean?" I asked.

"That could be the collapse," he replied.

A black hole.

Sarin had dozens of graphs to choose from, each telling the story of the last step of a binary neutron star dance billions of years ago. Some showed a lasting, stable emission of high-energy light for the duration of the observations. Survivors. Others seemed to hold on for up to a couple of minutes and then drop off. Some of them came together before Earth was even formed, their light finally arriving just as we developed the tools and talent to understand what we are seeing.

— The Last Straw —

Nine decades had passed since scientists had ascertained that there was only so much mass that can be piled onto the corpse of a low-mass star before the outward pressure of the electrons no longer supports the white dwarf's weight. Once it hits the Chandrasekhar limit and its electrons find refuge within its protons, the white dwarf collapses into a denser, smaller neutron star. It's the rare white dwarf that can actually achieve the Chandrasekhar limit, though, as they have a habit of blowing themselves to smithereens just before reaching that point. But the makeup of the core of a massive star does allow for this fate when it gets to the end of its last reserve fusion tank. The iron core begins collapsing into a neutron star when the Chandrasekhar limit is reached, the electrons unable to stand any further pressure. It's a feature that yields a fair degree of uniformity in the masses of neutron stars.

It's natural to then wonder: if there is a limit to how much pressure electrons can withstand, is there also a limit to how much pressure neutrons can cope with before they, too, buckle?

The answer is, apparently, yes. What is less certain is what the limiting mass is before a gravitational bottomless pit opens up in the fabric of spacetime.

"If you look back in the 1980s, people said the maximum mass of a neutron star should be maybe 1.8 or 1.9 solar masses," Sarin explained. "And everyone was happy. Then—whoops!—astronomers observed one with 1.98 times the mass of the Sun. So the nuclear physicists said, 'Let's try something new.'" The "something new" would have to allow neutrons to withstand more than twice the mass of the Sun squeezed into a city-sized ball.

Then a slightly more massive neutron star was discovered, and theorists were once again back at the drawing board. Astrophysicists at that point afforded the neutron star lots of wiggle room. Three times the mass of the Sun, they figured, was definitely, positively, absolutely the maximum mass that a neutron star could withstand before fully collapsing into something that creates such a steep dent in space-time that even light can't race up its falling sides and escape. As a result, that limit—three solar masses—shows up in some basic astronomy textbooks when the discussion turns to the fate of the most-massive stars. If the collapsing core has more mass, a black hole forms. Less than that? Well, let's just say that there's not a hard and fast limit like there is with a white dwarf.

So far, we have largely ignored stars capable of leaving behind black holes, but their story is essentially the same as their slightly lower-mass cousins that pack 9 to 25 times the Sun's mass. They will gorge themselves on hydrogen, fusing madly and producing the light of hundreds of thousands of Suns for a cosmic blink of an eye. This energy pushes outward, the atoms themselves acting as miniature sails catching the light and racing away from the gravitational grip of the star. A relentless light-driven wind moving at speeds of thousands of kilometers per hour gradually strips off the outer layers of these stars. The most-massive stars produce so much energy in their cores that they actually tear themselves apart, often expelling much of their outer layer of hydrogen in the process. The relatively lightweight Zeta Puppis, a blue supergiant with 25 times the Sun's mass, is currently losing about 1/500,000 of the mass of the Sun each year to these stellar zephyrs. But even when the stars tear themselves apart, the feast must go on. When the hydrogen is exhausted, the stars turn their ravenous appetites to helium, carbon, oxygen, and on down the line.

Most likely, such a star—at least to begin with—is in a multiple system, where it might merge with a companion as Eta Carinae did, a disruptive process that also helps the star to shed some unwanted pounds. Or perhaps the stars in a multiple system have enough distance between them that they each live out their lives held together by only a tenuous bond. No matter a massive star's path, its days are numbered, and once it fuses iron in its core, the party is over.

Does it go out with a bang? Possibly. The more massive member of Eta Carinae, with over 100 times the mass of the Sun, could ultimately wind up being something called a "superluminous supernova," a pinpoint of light that despite its enormous distance, would be brighter than the full Moon. It would be a formidable contender for the title held by the supernova of 1006, visible even during the day. When the dust ultimately cleared, the remaining object would not be a neutron star. The crushing pressure inside the heavyweight star of Eta Carinae would certainly breach the three-Sun limit in its core, so it would leave behind a black hole.

There's another way its future could play out, though, a scenario just as dramatic but significantly less flashy. The collapse that begins in the core simply might not end. For stars with less than about 25 times the mass of the Sun, the creation of neutrons provides a barrier for the inward rush of material trying to race down the gravitational drain. But Eta Carinae is far too massive. It's possible that the failure of electrons to support the iron core will be followed almost instantaneously by a similar failure of neutrons to support the crushing pressure. Anything falling inward would simply keep falling.

"In that case," Eta Carinae researcher Nathan Smith said, "you might just look out one day and say, 'Hey, Eta Carinae looks much fainter today' because you'd only be seeing light from the remaining companion star."

He continued: "The problem is that we can't really predict the end fate of some stars because we don't understand the physics well enough to say which stars of a certain mass will explode and which ones will collapse. But I'm hoping one of those things will happen to Eta Carinae right around the time I retire."

The lack of understanding is a problem. In fact, there isn't even a firm consensus on the lowest mass required for a star to leave behind a black hole. Some researchers push that figure as far down as 17 times the mass of the Sun, which would mean Betelgeuse, the star occupying Orion's armpit, might simply vanish as a black hole in the not-so-distant-but-probably-not-in-our-lifetime-and-certainly-not-before-Smith's-retirement future. So much depends on what is going on under the star's surface, including how quickly its various layers are rotating, the star's exact composition, and how much churning goes on inside as the star tries to pull energy out of its core with giant conveyor belts of unfathomably hot and violent plasma.

Even with all the uncertainties, astronomers think they have caught stars in the act of disappearing. For instance, in 2015, a red supergiant star in the galaxy NGC 6946, also known as the Fireworks Galaxy for its enthusiasm for producing supernovae, simply . . . vanished. Whether this disappearance was due to a complete collapse of the star or to the stellar equivalent of hiding behind a smoke screen is unclear. The Fireworks Galaxy is 25 million light-years away, so astronomers can't exactly jet over there to check on it. But the fact that such a collapse isn't out of the question reveals how far astronomers have come in accepting some of the completely bizarre things the universe conjures up.

— Clash of the Titans —

The universe doesn't care how a neutron star's weight limit is surpassed. All that matters is that it is surpassed—whether by the collapse of a too-massive core of a single star or by the merging of objects that individually are well below the maximum mass allowed for a neutron star. In the case of the Hulse-Taylor binary, which will come together in an upcoming geologic era (stay tuned!), the neutron stars weigh in at 1.44 and 1.39 times the mass of the Sun, each essentially at the Chandrasekhar limit. For now, these two are fine, but in 300 million years, they will merge to create something well over 2.5 times the mass of the Sun. When they do, the resulting object will likely be a black hole, but only after a fierce battle. As it turns out, it's one thing for a stellar core to implode under its own weight, but it's another thing entirely when two neutron stars experience a close gravitational embrace.

"It isn't a nice friendly handshake upon coming together. It's more of a ripping-apart explosion," astronomer Jeff Cooke explained animatedly, telling me this story as a prelude to another, seemingly unrelated one. "By the end, the intense gravity starts shredding them apart, and they start to get really stretched out, spilling neutrons everywhere. And that's messy, because if you have a neutron that's not inside an atom or in a neutron star, it only lasts for about 15 minutes."

That fact feels like it should get much higher billing in high school chemistry classes. Protons, the positively charged heavyweight cousins of neutrons, apparently have no expiration date, even when left out in the open. At least, that's what decades of underground Cherenkov detectors like Kamiokande II and its global siblings have told scientists. But making up a large percentage of the familiar material inventory of the universe are particles that, left to their own devices, will self-destruct within a lunch break. When two neutron stars come together, torrents of these particles are unleashed, and the neutron clocks start ticking. Once their internal timers run out, the neutrons spray the area with protons, electrons, and, to satisfy the universal accountant, the antimatter version of a neutrino.

Forcefully shredded from the gravitational prison of the neutron stars, the newly converted protons and escaped neutrons fuse to form atomic nuclei, the very things that the massive star's core squashed out of existence when it collapsed. In the ensuing mosh pit of protons, neutrons, and atomic nuclei, heavier and heavier elements are forged. It is all in a day's work for this stellar alchemist to whip up the Earth's weight in gold, and along with that comes a flood of radioactive elements that will gradually shed bits and pieces of themselves, along with gamma rays, for up to thousands of years.

The electromagnetic energy released from the initial flurry of destruction is considerable, perhaps a tenth of a foe, but this figure pales in comparison to the full foe that two white dwarfs conjure up upon colliding. But unlike a Type Ia supernova, which obliterates two white dwarfs in a stellar-mass fusion bomb, a merger like this blasts away only about 1% of the combined mass of the neutron stars. That's still a few thousand times the mass of Earth, but nothing compared to what is ejected in a supernova. As for what happens to the remaining 99% of the matter, well, that's the story that Sarin's graphs tell astronomers.

What his graphs don't reveal is their half-century-long backstory.

CHAPTER 14

Don't Blink

They are fast. Faster than you can believe.
Don't turn your back.
Don't look away.
—The Doctor, "Blink" episode of *Doctor Who*

— Perhaps a Bit Out of the Question —

Nikhil Sarin's work began where decades of hair-pulling frustration had ended.

"Most of the observations I interpret come from the NASA Swift Observatory or the Chandra X-ray Observatory," he explained. "Swift can see a good chunk of the sky at the same time. Once it detects a gamma ray burst, it automatically starts pointing its X-ray detector towards the source. This typically takes a few seconds."

"A few seconds," I thought in amazement. It has taken me longer to find the ~ key on my computer than it takes for this orbiting observatory to zero in on a cosmic flash. But I knew that things hadn't always been this straightforward.

His career just beginning when I spoke with him, Sarin was standing on the shoulders of giants like astronomer Kevin Hurley, who spent a good portion of his career chasing down the energetic fireflies first detected by the *Vela* satellites in the late 1960s. Hurley was just finishing his PhD when these gamma ray bursts (GRBs) burst onto the scene. Attracted by the high-energy unknown, he and others aimed to figure out what they could about flashes of gamma radiation lasting anywhere from milliseconds to tens of seconds.

The biggest problem then was localizing the bursts in the sky. GRBs pop off like paparazzi flashbulbs, never repeating.* Trying to determine exactly where they are is like trying to catch a firefly that flashes only once. Worse, an individual gamma ray detector at the time was good only for letting astronomers know *when* it had picked up a signal. If they wanted to know *where* an event had occurred, they needed several detectors spread out with as much space between them as possible.

So, in the late 1970s, Hurley spearheaded a plan to piggyback gamma ray detectors on spacecraft, the first of an armada of experiments that would become the Interplanetary Network (IPN). Soviet probes *Venera 11* and *12*, sent to explore the harsh environment of Venus, each took a French gamma ray experiment and trekked off toward our sister planet tens of millions of kilometers away. Closer to home was the *Prognoz* satellite, another Soviet spacecraft fitted with a French gamma ray detector. There was eventually a solar observer orbiting the Sun, a cometary explorer far above Earth's surface, and a mission to Mars. Basically, any spacecraft that could accommodate a gamma ray sensor got one.

The vast distances between the spacecraft made the IPN a powerful tool in narrowing down the direction of the bursts. Just as a clap of thunder isn't heard by everyone at the same time, a gamma ray burst isn't detected by every station simultaneously. If a spacecraft orbiting Earth picked up the signal first and one near Mars caught it a few minutes later, then scientists knew that the basic geometry was "source, then Earth, then Mars." Adding more spacecraft helped localize the source even better. This multinational ensemble gave astronomers a fairly good idea which direction a burst came from, even if there was no way to triangulate the distance.

The information was far from instantaneous, though. It took a day or more to get the information that gamma rays exceeding a predetermined threshold had been detected by spacecraft at various locations in the solar system. By the time astronomers sorted out the general

---

* Well, maybe mostly never repeating. A handful of even more bizarre objects have turned out to be in a class by themselves

patch of sky that had hosted the flash, the event was long gone. But that shouldn't matter, they figured. Surely, something would remain, and all they would have to do is point an optical telescope at the location of the burst, find something obvious, and, voila! The mystery would be solved, and they could move on to the next puzzle.

When we spoke, Hurley recalled, "We naïvely thought that the sources would be there for a long time, and you could just go back and look a month or two later, and you would see something."

When the hunt for GRBs began, everyone's money was on a straightforward phenomenon, perhaps the neutron star version of a nova. A companion star would overflow its territory, and then its material would race down the steep gravitational pit in spacetime where, well, um . . .

The exact mechanics of producing an epic burst of gamma rays from a neutron star siphoning matter from its partner were unclear, but that was something that could be sorted out once an optical follow-up observation had been made. "But as we accumulated more and more of these positions, we kept turning up—" Hurley paused. "Absolutely nothing."

No remnants. No lingering glows. No further clues.

He continued, "Then we thought maybe we just needed to work faster, and the race was on to build instruments that could get rapid locations." That race turned from a sprint into a marathon, continuing into the 1990s and beyond as astronomers fought to catch GRBs in the act.

It was absolutely imperative that they get at least approximate locations, down to a few degrees at worst, and then plot the GRBs on a map of the sky. Even though it wouldn't be perfect, it would at least help astronomers determine whether the bursts were coming from inside our Milky Way Galaxy or beyond. If the GRBs tended to fall along the plane of the Milky Way Galaxy, that cloudy ribbon of sky observable far from the glow of city lights, then they were most likely relatively nearby, which, to an astronomer, is just thousands to tens of thousands of light-years away. If the bursts happened all over the sky, then there were two wildly different possibilities: They were either very close to the Sun, scattered less than a few hundred light-years away, or they were in distant galaxies perhaps billions of light-years away.

The arguments in favor of local sources were the same ones that had been trotted out for a century. Nearby sources, while still pumping out the equivalent of a year of sunshine in seconds, would require the mysterious objects to generate energies that made sense. Accepting that GRBs were coming from well outside our own galaxy would mean accepting that something in the universe could spit out a foe of energy in the snap of a finger. In gamma rays, no less. The sheer power required for such a feat turned many astronomers away from the idea that they could be outside our Galaxy, but until more definitive data came in, there was no real evidence either way.

By the early 1980s, with over 100 discoveries and 46 general locations ascertained, astronomers announced that GRBs seemed to be happening randomly all over the sky. There was still a glimmer of hope that they were extremely close, a conclusion that suited 95% of GRB researchers just fine.

Hurley recalled the uncertainty. "I thought every Monday, Wednesday, and Friday that gamma ray bursts would turn out to be some sort of Galactic phenomenon related to neutron stars, and then on Tuesdays and Thursdays, I thought, wow, what if they were extragalactic? Wouldn't that be cool?"

"What about weekends?" I asked.

Hurley laughed. "On Saturdays and Sundays, I just didn't think about it at all because it was so confusing."

Amid all the confounding cosmic paparazzi, the last thing they needed was GRB 790305. Following the naming convention of GRB for a gamma ray burst and then year-month-day, it was affectionately known as MF.

For March Fifth, of course.

— Speak Softly and Carry a Big Energy —

It probably wasn't even a true GRB. The tsunami of gamma rays that washed over the solar system on 5 March 1979 was an event that stood out even among the standouts. It was announced in May of that year through the International Astronomical Union's Central Bureau for Astronomical Telegrams (CBAT), the same channel that had announced countless novae and supernovae. First, the Soviet probes

Venera 11 and *12* had spiked, and within minutes, ten other missions in the Interplanetary Network found their detectors saturated. With such a brief blast and so many detections, astronomers were able to pin down the source not just to within a couple of degrees in the sky, but to a region only one-sixtieth of a degree across. Within that box, deep in the Large Magellanic Cloud, was a relatively young supernova remnant discovered by radio astronomers. It was a toddler, only a few thousand years old.

There was no way, scientists said, that this was just a coincidence. But there was also no way that they could account for so much energy emitted by an object 180,000 light-years away. Within a tenth of a second, it had radiated the same amount of light that the Sun produces in a millennium.

It wasn't finished either. Less intense gamma rays pulsed every eight seconds for over a minute after the initial blast, and as the years passed, the object that had produced GRB 790305 occasionally and without warning belched out lower-energy gamma rays. The periodic gamma rays following the initial intense burst were fine, astronomers conceded, but anything beyond that was simply unacceptable. GRBs might be enigmatic and inexplicable, but with a growing catalog of known events, astronomers were certain that they were one-and-done.

And maybe GRB 790305 was actually done, at least as a gamma ray burst. The signals that came in later fell in a regime known as "soft gamma rays." X-rays, really, at the low-energy portion of the gamma ray spectrum. The March burst itself not only was more intense, but also consisted of higher-energy "hard gamma ray" photons, not the later soft ones. Don't let the name fool you, though. Even a soft gamma ray photon is tens of thousands of times more energetic than a visible photon, so there is nothing cuddly about an object that emits a sudden pulse of this type of light.

Conveniently, soft gamma repeaters (SGRs), as they became known, are recurrent. Not on a regular time scale—that would be too convenient—but these fireflies do blink more than once, a property that has allowed astronomers to catch them in an ever more certain jar. Resistance to the idea that the March 1979 event had been as far away as the Large Magellanic Cloud gradually—*very* gradually—dissolved.

The year 1979 saw the discovery of another soft gamma repeater, one that flashed over 100 times in as many months with varying intensity and absolutely no rhyme or reason to its activity. With so many bursts, SGR 1806-20 practically begged astronomers to pinpoint its location: near a supernova remnant over 40,000 light-years away. After years of chasing these particular fireflies, astronomers were able to definitively associate the three then-known SGRs with neutron stars, but these weren't run-of-the-mill neutron stars.

Run-of-the-mill neutron stars . . . as if such a thing exists. It seems unfathomable that our explorations into the gamma ray universe had pushed astronomy to the point that neutron stars could ever be considered ordinary. They are stellar masses collapsed to densities beyond imagination, some spinning faster than a kitchen blender. They're armed with magnetic fields that can shred you and put one of your charged particles on one side of the room and another on the other side of the room.

These things are anything but ordinary.

Soft gamma repeaters, though, implied something even more extreme. The theory was fuzzy, but they seemed to be newborn neutron stars trying to relieve the tension in their immensely powerful and tangled magnetic fields. They are magnetars, and the fact that there are definitely some of these in our Galaxy would eventually become cause for concern.

But what about the other gamma ray monsters? Where are they?

— Bursting with Energy —

Observers weren't the only ones trying to wrap their brains around the energetic new twist the universe had thrown our way. Theorists were also scratching their heads as they tried to understand what could produce such a burst of gamma rays at any distance. Frustratingly, other than being bursts of gamma rays, no two GRBs appeared to be the same. The puzzle pieces were scant, but as in the case of Bell Burnell's original pulsar, timing was everything. Although some GRBs lingered for over a minute—these were cleverly designated "long bursts"—some winked in and out in mere milliseconds. Those were the short bursts, and astronomers figured that their underlying mechanism had to be different from that of their long burst brethren.

Long or short, GRBs showed tiny, even shorter-term variations in their signals, variations that implied that the source was very small. Very small indeed, and capable of generating gamma rays without obviously emitting any of the lower-energy types of light (and certainly not in any lingering fashion, if they did). The obvious suspects were neutron stars and black holes, objects that had been purely hypothetical just a generation before. If anything could pull off extreme events, they could.

The impending merger of the Hulse-Taylor binary neutron star provided another possible explanation. By the early 1990s, astronomers had observed a whopping four pulsars in binary neutron star systems in our Galaxy. Fortunately, the universe is brimming with galaxies, and those galaxies are brimming with stars, and many of those are massive stars in binary systems. Massive stars in binary systems live fast and die hard, and after their explosive deaths, they have a fighting chance of leaving behind a binary neutron star system. Even from this paucity of nearby data points, astronomers estimated that there would be a binary neutron star merging event within a few billion light-years about once every three days. The entire observable universe would host a handful of these mergers each day.

But how to observe such a thing? During a neutron star merging event, the vast majority of the energy—over 100 foes of it—is blasted out as neutrinos and gravitational waves. The gravitational waves would be the smoking gun, but in 1991, gravitational wave astronomy was still a dream. The US Congress had approved funding for two ambitious gravitational wave observatories, but the sites hadn't even been determined yet. It would be at least a decade before either was up and running, and easily another decade before they were fine-tuned to the point of picking up wiggles in spacetime that are characteristic of two colliding neutron stars.

Gravitational wave detection was clearly out, but what about neutrinos? Surely, SN 1987A had proven that we could catch those. True, but SN 1987A was practically in the same room with us, and we caught only two dozen of its trillion quadrillion quadrillion quadrillion neutrinos. A collision between two neutron stars 10,000 times farther away would give us nothing.

Thankfully, the universe has always been involved in countless cosmic energy laundering schemes, and theorists posited that there was at least one channel that might give us a window on a neutron star merger. If a flood of neutrinos and their antimatter counterparts (antineutrinos) converted to electrons and *their* antimatter counterparts (positrons), then the electrons and positrons would meet up, annihilate each other, and create gamma rays. This chain of events is not as preposterous as it may seem at first glance. As long as the universal accountant is satisfied, swapping out one type of particle for another or even cashing in particles for photons is not only permissible, but commonplace in the most extreme environments in the universe. A swift burst of gamma rays could indeed be the most easily observed consequence of a neutron star merger. It was a long shot, particularly without any additional way of confirming the event, but it did provide one possible answer.

On the other hand, a different cohort of astronomers argued, supernovae were far more common, so perhaps they were the bangs behind the bursts. Perhaps there was something about the exact geometry of the collapsing core and the subsequent explosion that shot out gamma rays in narrow jets, like cosmic blowtorches. As in the case of pulsar observations, if Earth happened to stare down the barrel of one of these jets, we would see a burst. If not, well, we'd miss out, just as we miss out on countless neutron stars that don't point their beams of energy our way.

On another hand—you'll need to start using your neighbor's hands—maybe these are just bizarre things in the outskirts of our own solar system. On yet another hand, perhaps there is something far more exotic going on, completely beyond the scope of 1990s astrophysical understanding. Or perhaps, more than one answer is right. Maybe the shorter bursts come from one type of source, and the longer bursts come from another.

There were plenty of hypotheses. Legend has it that at one astronomical meeting on GRBs, a speaker claimed that it was harder to find someone who hadn't proposed an origin for GRBs than to find someone who had.

So much had changed since the appearance of a new dot in the Andromeda Galaxy a century earlier, and yet so much remained the same. Answering the question required more data, and there was a new urgency. They needed some way to instantly follow up on a burst detection.

— Location, Location, Location —

Getting a detailed picture of the universe in gamma rays was not nearly as straightforward as, say, mapping the visible or radio universe. Astronomers wanting to see in other wavelength regimes need orbiting observatories, and as a rule, those are expensive, painstakingly difficult to construct, and fraught with peril.* A launchpad mishap can turn years of planning and work to debris, and remedying a tiny miscalculation in design can be challenging at best, impossible at worst.

Still, NASA agreed that exploring the multiwavelength universe was the only way to get the answers to the biggest questions, and it committed billions of dollars to its four great observatories to do just that. The first, launched in 1990, was the Hubble Space Telescope, which would—after a servicing mission to correct its slightly astigmatic optics—give us the clearest view yet of cosmic goings-on. Then, just a year after the Hubble Space Telescope was sent into orbit, the Compton Gamma Ray Observatory (CGRO) hitched a ride on the Space Shuttle *Atlantis*, from which it was gently sent into orbit. The year 1999 saw the launch of the Chandra X-ray Observatory, and in 2003 the Spitzer Space Telescope was launched to reveal the infrared universe in unprecedented detail.

Occupying the eight corners of the CGRO was BATSE—the Burst and Transient Source Experiment—which was able to view the entire gamma ray sky 24/7. Immediately, it began picking up one GRB per

---

* Ground-based observatories also have their fair share of peril. The 2.7-meter Harlan J. Smith Telescope at the McDonald Observatory in Texas has several bullet holes in its primary mirror from the gun of a disgruntled employee.

day. They were everywhere. Up. Down. Left. Right. After several hundred had been logged and analyzed, astronomer Chryssa Kouveliotou grudgingly reported that "a consensus is slowly forming within the community toward an extragalactic origin."

There was still no definitive proof, though, no clear evidence that any particular GRB occurred in any particular galaxy. Then in 1996, an Italian-Dutch X-ray and gamma ray observatory called BeppoSAX was launched, and just ten months later, astronomers finally got what they had been after for three decades: the precise location of a gamma ray burst. Astronomers celebrated this milestone, but they were billions of years late to the party. The light from GRB 970228 had been traveling for more than half the age of the universe by the time the blast swept over our instruments.

"When the first one came along that had a pretty secure redshift, that was a really nice moment," Hurley recalled. "It made it all worthwhile."

Simon Johnston, a senior research scientist at CSIRO, had a slightly different take on the first definitive host galaxy. "As soon as they had localization, it was game over. Gamma ray bursts were extragalactic." He paused thoughtfully. "But they're not actually useful for all that much in terms of cosmology."

"Useful" was not a word I expected to hear about something billions of light-years away, but it did reveal a fascinating divide in astronomers. Some are excited about the extreme events themselves. I mean, what could be more thrilling than crushing gravity, hyperactive spin rates, immense blasts of energy, and distortions in spacetime? Other astronomers, while still suitably in awe of the objects they study, are more interested in using them as tools to understand the grander goings-on in the universe. I recalled my chat with Suntzeff, who had been using Type Ia supernovae to chase H_0 and q_0, the numbers revealing how the universe itself is behaving. His interest in the supernovae is secondary to their utility.

And it wasn't clear if GRBs were anything other than cosmic show-offs.

— The Jet Set —

Useful or not, these beasts can pack a punch. After BeppoSAX's first localization, which proved that at least some GRBs are indeed at truly cosmological distances, billions of light-years away, astronomers could then begin to sort out the problem of their power source, and it was definitely a problem. If researchers assumed that the same amount of energy went out in all directions, the bursts were clearly out of the question. There simply was no way for the universe to pull off this feat. Full stop. Yes, this seems like a very familiar story, but this time they were *really* serious.

Something that the universe had revealed to us for decades, though, is its talent for channeling energy and particles into narrow beams or jets. There are pulsars, after all, and that odd tail jutting out of the quasar 3C 273, but there are plenty more. Jets had been first observed in 1918 by Heber Curtis, who noticed a "curious straight ray" in the galaxy M87. That curiosity turned out to be a 5,000-light-year-long stream of high-energy particles racing away from the heart of this otherwise unassuming galaxy. On a much smaller scale, stars in the process of forming—protostars—also kick out jets.

Making a jet is, it turns out, a fairly universal process wherever gravity is involved. Material falling toward a massive object never seems to make a direct hit, but instead forms a hot, turbulent disk that spirals inward. It's the same reason a pancake of matter forms around a white dwarf as it siphons material off its bloated companion. As that matter jostles toward its gravitational destination, it becomes hotter, and the atoms are stripped of their electrons. With so many charged particles zipping in tighter circles, the disk essentially becomes an enormous and enormously strong electromagnet. Caught in the ever-tightening grip of some of the strongest magnetic fields the universe can muster, particles shoot out in tight columns above and below the disk. Think Saturn, but with blowtorches blasting out of its north and south poles.

With its energy concentrated into two narrow cones, a jet-producing object gives different observers distinctly different views. If you imagine staring down the barrel of one of these jets, it's easy to see how you might be fooled into thinking that the object is far more powerful than it is. If you observe it from the side, though, it doesn't seem so impressive. In fact, if an object isn't firing off its jets in your direction, it might not even be visible. Like the pulsars that never swing their beams past us, these are probably the norm, rather than the exception.

Having netted a host of jetted objects in the universe by the 1990s, it wasn't much of a leap to blame jets for gamma ray bursts. For one thing, this solved the energy crisis.

"The energies needed are, maybe, 10^{52} ergs," Hurley explained nonchalantly.

Oh. "Only tens of foes," I thought. Whatever.

Sensing my disbelief, he added, "That's *maybe* a hypernova, but it isn't requiring us to invoke any kind of new physics or exotic types of stars."

Readily accepting this kind of energy output is major progress for a profession that once refused to admit that the Sun's entire lifetime energy output could be compressed into months.

As more and more of these objects popped off, and as astronomers got better at catching them in the act, two prevailing candidates for GRBs were, if not universally accepted, at least most plausible.* Shorter bursts, which make up only about a third of the total, seemed to be best explained by neutron star merging events, an explanation that was just the starting point for Sarin's research in the early 2020s. If the two neutron stars are comparative lightweights, the resulting object survives. If not, a short-duration GRB might be the universe's way of telling astronomers that it has just cooked up a brand-new black hole.

Astronomers generally agree that longer-duration GRBs always herald the formation of a black hole. To produce these bursts, the cores of some of the most-massive stars collapse. Those cores spin faster and faster and faster still, disks whirling at relativistic speeds as the center of the star tries to race down the newly unplugged gravita-

* These two explanations, of course, utterly ignore the magnetar flare explanation for the SGRs, which are an entirely different puzzle.

tional drain. From the poles come powerful jets that blast through the outer shell of the star like bullets.

From a visual standpoint, events responsible for the long-duration bursts would also yield some of the brightest supernovae, objects known as superluminous supernovae or hypernovae.* Eta Carinae might very well become one of these, or it might simply collapse. In either event, it will most likely fire out jets along its poles, blasting away much of the core-collapse energy into beams of high-energy radiation and fast-moving particles that, thankfully, will not be pointing in our direction. Astronomers have already seen matter ejected at Eta Carinae's poles, and they are safely aimed away from Earth.

But what if they're wrong?

---

* At times, it seems that astronomers have run out of superlatives to describe what goes on in the universe we inhabit.

CHAPTER 15

Point Blank

The interesting thing about staring down a gun barrel is how small the hole is where the bullet comes out, yet what a big difference it would make in your social schedule.

—P. J. O'ROURKE, *HOLIDAYS IN HELL*

— Ready, Aim . . . —

Of course, astronomers have tried to figure out exactly how the story would play out if Eta Carinae had us in its sights, just in case. Spoiler: Nobody, not even the dog, survives. From a mere 7,500 light-years away, give or take, a foe of gamma ray energy would destroy our ozone layer and sandblast the surface of our planet and all the life on it with bizarre particles called muons. They are forged in the upper atmosphere from the energetic collisions between high-speed cosmic particles, gamma rays, and the usual constituents of air, and there is a constant drizzle of about a muon per minute hitting you even as you read this. It's a type of radiation that human physiology has learned to live with. A gamma ray burst from Eta Carinae or another relatively nearby source, though, would turn that muon drizzle into a deluge. Between that and the ensuing flood of high-energy radiation that our now-absent ozone layer protected us from, life on Earth would be obliterated.

It's worth repeating at this point that the poles and equator of the Eta Carinae system are well known, and we are fortunately not staring down the barrel of that particular gun.

There are, however, other guns.

Take Apep, for example.

"It's named after the ancient Egyptian god of chaos," explained Benjamin Pope, whose research interests involve developing mathematical techniques to explore our universe. Because these tools can be applied broadly, he has found himself studying everything from Eta-Carinae-like objects to tree rings.

He pulled up the Wikipedia page for "Apep (Star System)," not to be confused with Apep, serpent deity and mortal enemy of the Sun god Ra, on his phone. "This thing was discovered by Joe Callingham back in 2019," he said as he tapped on the image. It looked strikingly similar to the binary star windmill animation that Orsola De Marco had shown me.

"You've got three stars in here," he explained. "There's a companion that we don't think is interacting with the relevant physics very much, and then the bright one in the middle is a binary."

That's why it looked so familiar. And yet, despite being a familiar binary system spewing out a pinwheel of hot gas, Apep is very, very different.

"It's the only known binary of two Wolf-Rayet stars," Pope continued. "These are two massive stars at the end of their lives."

Wolf-Rayet stars are leviathans, so massive and generating so much energy in their cores that they are blowing off their own outer layers at speeds of thousands of kilometers per second. These stellar gales can strip away many times the mass of the Sun, exposing the layers beneath the hydrogen and setting up the stars to become Type I, but not Type Ia, supernovae.

"The winds from the two stars collide, and then when that collision orbits around, it's like a sprinkler," Pope explained. "It produces an Archimedean spiral—this beautiful spiral—that wraps around."

The fact that astronomers are seeing a beautiful spiral pinwheel means, unsettlingly, that we are looking down on this system from above, potentially right above the axis where any energetic jets will shoot out. More unsettling is that Apep is not the only such pinwheel that astronomers have discovered in our Galaxy. WR (for Wolf-Rayet) 104 is another, and both it and Apep are well within the distance that could trigger a mass extinction on Earth should they (1) become long gamma ray bursts and (2) have us in their sights.

Whether either would create a GRB event and whether that event would in fact point its jets our way are far from certain. GRBs in our current universe are harder to make than they were in the past. The decline is attributed to the fact that today's stars contain yesterday's stars' products. As the concentration of elements heavier than helium grows, the conditions for long-duration GRBs become less favorable. We believe that Apep and WR 104 got that memo, but it's hard to tell through all the intervening dust, which obscures our view of them.

Still, Pope cautioned against sensationalizing these objects, recalling that when the original news of WR 104 hit the public in 2008, the object found its way into countless articles and documentaries.

"They wanted to know, could it blow up Earth?" Pope said of one of these documentaries. "And the astronomers were like, '*If* it goes off, and *if* it has this rapid rotation, there's every chance it would be a gamma ray burst. We have no reason to think the core is rapidly rotating, so it wouldn't produce a GRB. Just a supernova.' And then there's this great voiceover: 'But what if they're wrong?'"

This kind of overwrought dramatization catapulted WR 104 into countless end-of-world theories surrounding the 2012 Mayan calendar. When the world didn't, in fact, end in December of that year, the interest of the layperson waned, but astronomers were still keen to better understand the system and perhaps get a more definitive answer to whether life on Earth will be extinguished by something beyond our control.

Peter Tuthill, an astronomer who has been at the forefront of researching WR 104, has this to say on his website:

> There are a lot of missing or incompletely understood links in the chain of logic connecting WR 104 to a real threat to Earth. But conversely, I don't know of any showstopper that categorically rules it out either. I think it is an obligation for scientists to keep an open and impartial mind to any event that could have consequences to a large number of people, and this would appear to be one of them, even if the odds appear quite long at the present time.

What we do know is that such systems are incredibly rare. The entire Milky Way Galaxy, with several hundred billion stars, is home to just a handful.

That might be all it takes, though. Some astronomers estimate that every few hundred million years, there *is* a gamma ray burst that fires at us from the cosmic equivalent of point-blank range. This statistic, along with evidence in Earth's rocks of an enduring period of ultraviolet light, suggests that 360 million years ago, a GRB stripped Earth of its ozone layer and prompted a mass extinction event. Again, the data are far from conclusive, but it is not out of the question that gamma ray bursts have directed the development of life on Earth. If that is the case, they have been frighteningly useful objects, even if cosmologists disagree.

— A Barrage of Bullets —

Having exhausted the topics of gamma ray bursts, binary Wolf-Rayet stars, and Mayan calendar end-of-days conspiracies, Pope asked the natural follow-up question: "So what do you know about dendro-chronology?"

I knew more about that topic than I did about how it fit into our present conversation. For instance, I knew that scientists can divine quite a bit about the life of a tree by looking at its rings. Drought years reveal themselves as narrower rings than wetter years, so by counting back from the outermost ring, dendrochronologists can determine something about each year in the life of a tree. What I didn't know was that by piecing together the lives of innumerable species of trees that were felled at different times in different places, researchers had managed to amass entire libraries of tree rings dating all the way back to the last Ice Age. This unbroken lineage gives them the ability to pinpoint historical events with single-year accuracy—and cosmic events as well.

Pope was almost reverent about the subject. "Trees provide this world-scale observatory. There's information that has been hiding all around us all this time. It's amazing that if you go and touch a bristlecone pine, in its body—encoded in ways that we can't see—are measures of extreme astrophysical events."

I didn't at first understand what he meant, but I would soon appreciate the unusual and unlikely observatories growing all around us.

Their story begins with carbon. Carbon is distinctly different from other elements because it contains exactly six protons. There are different isotopes of carbon, though, and those contain different numbers of neutrons. Carbon 12 has six neutrons to go with its six protons—12 total baryons—and is the most stable of the carbon isotopes, its nucleons never seeing the need to convert or to depart the nucleus. The story is a bit different for carbon 14. Over the next 5,730 years, which is the half-life for this isotope, there is a 50–50 chance that one of the eight neutrons in a given atom of carbon 14 will spit out an electron and an antineutrino (to satisfy the universal accountant) and become a proton. If it does, the atom becomes nitrogen 14, which contains seven protons and seven neutrons.

Given that our planet has been around about a million times longer than a carbon-14 half-life, it might seem surprising that any carbon 14 exists in our atmosphere at all. But the atmosphere is subjected to a continual bombardment by high-energy particles—mostly protons—known as "cosmic rays," some of which originate in our humble star and some of which are hurled out of the most distant and violent regions of the universe. An unsuspecting atom in the path of one of these cosmic bullets is shattered, creating a shower of both familiar and exotic particles (e.g., muons) that rain down on our planet.

The term "cosmic bullets" is not as hyperbolic as it sounds. The most energetic cosmic ray particle detected literally packed the energy of a 100-kilometer-per-hour baseball. It was a single proton traveling at the seemingly impossible clip of 99.9999999999999999951% of the speed of light, earning it the semiofficial designation of the Oh-My-God (OMG) particle. When this proton crashed into our atmosphere in 1991, it released a torrent of particles that released additional torrents of particles that caused the nitrogen in a patch of atmosphere to glow.

This air fluorescence caught the attention of the unusual Fly's Eye Observatory in Utah. Like neutrino detectors, the Fly's Eye was designed to see not the particles of interest, but rather what they do to the neighborhood as they come crashing through. In the case of cosmic ray observations, our atmosphere is that neighborhood. When suitably energized, it yields an observable faint blue glow. In the three decades following the discovery of the original OMG particle, scientists have spotted only a handful of similarly energetic cosmic rays. Unfortunately, there are still no solid leads on where these things come from. The usual suspects—supernovae, gamma ray bursts, even quasars—all seem to fall short.

— Rings of Power —

A blue glow is the most immediately observable consequence of cosmic rays in our atmosphere, but it's not the only consequence. Cosmic rays also induce a steady supply of free-roaming neutrons in the upper atmosphere, particles that will self-destruct in minutes if left in the open. When a neutron of suitable energy interacts with a nitrogen-14 atom, it can evict a proton and settle in with its new family of carbon-14 nucleons. Given that most of the atmosphere is nitrogen, this particular reaction is relatively common and allows the atmospheric level of carbon 14 to be constantly replenished.

Meanwhile, living things continually take in carbon from the atmosphere (or from other living things that do so), and since the carbon-14 content of the atmosphere is relatively stable, so is the carbon-14 content of a living thing. When something dies, it stops taking in carbon 14, and any carbon 14 already present in the body will slowly decay into nitrogen 14. Eventually, after many, many carbon-14 half-lives, there will be no measurable carbon 14 left, but until that point, the proportion of carbon 14 can be read like a clock. Since the late 1940s, scientists have used this statistical behavior of carbon 14 to determine the ages of archaeological objects.

There are, of course, complications. While it's true that the proportion of carbon 14 in the atmosphere is reasonably steady, it's not perfectly so. If the amount of carbon 12 goes up because of, say, a volcanic eruption, the hands of the clock get jostled a bit. To get truly accurate radiocarbon dates, researchers have to know how much time the clock has gained or lost and when.

Which brings us back to tree rings. In 2012, physicist Fusa Miyake at Nagoya University was studying Japanese cedar trees and found a sudden spike in the fraction of carbon 14 in rings accurately dated to 774 CE. This was not some mild blip in the data, but a factor of

20 times the sort of increase correlated to the things the Sun is known to do. Further scrutiny of other tree ring samples yielded several similar events marked by a sharp increase in carbon 14 followed by a years-long gradual reduction as the carbon 14 became largely sequestered in Earth's oceans.

Trees don't lie, and what they are telling scientists is that once every 1,500 years, give or take a few centuries, something somewhere suddenly and ferociously blasts our upper atmosphere with intense energy.

"That's all we have." Pope shrugged. "We don't know where they [come] from. We don't know what they are."

Like someone looking for lost car keys under the brightest streetlamp, scientists then determined that the best place to start looking was right next door—at our humble and mostly well-behaved Sun.

— Forecast: Solar Storm Ahead —

The Sun's good behavior, however, is an illusion. In 2020, NASA and the European Space Agency compiled a video showing 25 years of data from the Solar and Heliospheric Observatory, a spacecraft that has been doing what parents everywhere tell their children not to do: stare at the Sun. With each day compressed into a fraction of a second in this video, you become acutely aware that the steadily shining Sun is a constantly sparkling buzz of activity. An enormous plasma bubble shoots out here, another is coughed up there. Occasionally an energetic blast washes over the spacecraft, and the image briefly turns to snowy static as the instruments are momentarily overwhelmed by the charged particles racing off the Sun's face.

The movie is simultaneously hypnotic and sobering. A single modest eruption from the surface of the Sun can shoot out unfathomably hot plasma into space at thousands of miles per hour. Watch the video for just a few seconds, and you realize how unsettlingly frequent such occurrences are. The energy to produce these flares and coronal mass ejections is peanuts, of course, compared to the seething power deep beneath the Sun's surface, which is handily contained.

The worst of the solar flares might launch a mere 10^{-19} foes in a minute. In the grand saga of the energetic universe, this minor burp hardly merits a footnote except for the fact that its power source is eight light-minutes away. A millionth of a trillionth of a foe at that distance distracts one's attention from the foe per second that Apep or WR 104, both thousands of light-years away, might hypothetically shoot our way in the nebulous future. But neither of them is known ever to have committed any crimes against us.

The Sun, however, has a fairly lengthy rap sheet.

"While engaged in the forenoon of Thursday, September 1, in taking my customary observation of the forms and positions of the solar spots, an appearance was witnessed which I believe to be exceedingly rare," wrote Richard Carrington in 1859. "Two patches of intensely bright and white light broke out. . . . My first impression was that by some chance a ray of light had penetrated a hole in the screen attached to the object glass, for the brilliancy was fully equal to that of direct sun-light."

What later became known as the "Carrington event" was a spectacular solar explosion. The bright white flash that he saw was a flare that despite being so obvious visibly, released most of its energy in X-rays and gamma rays. Thankfully, this high-energy light was stopped by our upper atmosphere, but the Sun had another surprise in store for us. The continual churning of the interior had caused magnetic fields to twist, writhe, and ultimately snap like overstressed rubber bands. The release had blasted out a missile of hot plasma, and our planet was squarely in its cross hairs.

Seventeen hours later, that missile slammed into our planet, temporarily weakening the shield of our magnetic field. High-energy charged particles managed to burrow more deeply than usual into Earth's magnetic bubble and illuminated our upper atmosphere. The aurora borealis, typically seen only at arctic latitudes, where energetic protons and electrons are most often funneled, was seen as far south as the Caribbean. Telegraph operators received literally shocking news, and some could transmit and receive messages even after having turned off their power supplies.

Carrington was correct in his assessment that this type of activity was "exceedingly rare." The 1859 explosion was a solar superstorm, an event estimated to affect Earth less than once per century, even though the Sun is frequently coughing up such things. Slightly less intense storms find their way to our neighborhood every few decades, and the more we rely on electrical grids and electronics, the more these tempests impact our lives. In 1921, a blast of solar plasma disrupted telephones and telegraphs and even sparked fires. In 1989, the entire Canadian Hydro-Québec electricity transmission grid crashed, and the power supply to a nuclear power plant was damaged. Halloween in 2003 saw the Sun produce an enormously bright solar flare and subsequently send a wash of hot plasma Earthward, causing localized blackouts and prompting airlines to alter their usual polar routes. So much energy was pumped into Earth's atmosphere that it briefly expanded, tugging at satellites in low-Earth orbits. For a few days, nobody could pinpoint the locations of nearly half the satellites then in orbit.

And these were not even solar superstorms.

In fact, neither they nor the actual Carrington event could have produced the spike in carbon 14 seen in the tree rings.

"The particle fluxes implied here suggest, maybe, 80 times the Carrington event," Pope revealed.

"Holy shit!" I exclaimed.

"Yeah. It would take out every satellite and even undersea cables. We would lose the internet and intercontinental communication." He added emphatically, "There's a 1% chance in the next decade that, by surprise, we will lose our entire technological civilization."

And we may lose it to the same object that has allowed that civilization to thrive, the ultimate bait-and-switch. "But it might not even be the Sun," Pope said encouragingly.

Written records from the past millennia abound. Given that a super–Carrington event would generate spectacular auroras in skies that don't typically host them, it seems that someone somewhere would have mentioned an eerily glowing sky. One such report might be found in the *Anglo-Saxon Chronicle*, where there is mention of a "red crucifix after sunset" in 774 CE, an apparition that might have been an aurora or might have been a supernova obscured by dust. Like so many ancient texts, this entry is somewhat inscrutable.

Still, . . . the *Sun*? It might be a bit boisterous, but surely it can't be capable of this sort of devastation, can it? And it's not just astronomers saying that they can't believe it would behave so badly. Tens of thousands of Sun-like stars have been observed for years, notably by the Kepler Space Telescope and more recently by the Transiting Exoplanet Survey Satellite. With the rarest of exceptions, middle-aged, slowly rotating stars like the Sun simply do not spit out the energy equivalent of 100 Carrington events.

On the other hand, plenty of stars—usually younger, usually cooler, and usually spinning faster—have been found to be capable of that level of destruction and more. Prospects for life on any planet around the star AD Leonis were effectively shot in 2020 when astronomers witnessed a superflare the first night they observed it. The most intense stellar superflares can blast off as much energy as a million Carrington events. Although only emitting a millionth of a billionth of a foe, these superflares are enough to sterilize a star's planetary system.

A few dozen Carrington events of energy would be bad enough, but that's thankfully as much as the Sun will ever manage. And there is the possibility that it's not the Sun causing the carbon-14 spikes. Over the past few decades, astronomers have come to grips with the discomfiting reality that even enormously distant objects can affect us.

Objects like SGR 1806-20.

— Magnetic Krakatoa —

It would seem at first glance that 42,000 light-years would be a sufficient buffer zone between us and anything that the universe could throw our way. Overkill, really. The distance from the solar system to the center of the Milky Way Galaxy is less than that, and astronomers are confident that the black hole containing 4 million Suns of matter lurking in our Galaxy's heart is not a threat. But more than 10,000 light-years on the other side of that monster lies a tiny dot known as SGR 1806-20.

On 27 December 2004, minus about 42,000 years for light-travel time, a dot that had been quietly but persistently begging for our attention since 1979, firing off soft gamma rays on an irregular basis, decided to throw a tantrum. Fortunately for astronomers, the Swift

Gamma-Ray Burst Explorer had just been launched a month earlier. Three frustrating decades of uncertainty about GRBs had been just too much, and astronomers had pulled out all the stops with this orbiting observatory. Armed with a gamma ray burst alert telescope, an X-ray telescope, and an optical telescope, Swift—which would later be renamed the Neil Gehrels Swift Observatory—had the ability not just to notice that a gamma ray burst had tapped on its shoulder, but to whip its other telescopes around in moments to catch the tapper in the act.

But SGR 1306-20 hadn't given us a playful tap. Instead, it had hit Swift in the back of the head with a brick. The gamma ray sensors were completely overwhelmed for a fraction of a second as a tsunami of high-energy photons raced past. Optical, X-ray, and even radio follow-up observations all pointed to the attention-seeking soft gamma repeater. Astronomers were dumbfounded. Sure, it had thrown out volleys of X-rays and gamma rays here and there, but it had never shown a propensity for this kind of violence.

In a tenth of a second, it had released a whopping 0.00000000001 foes of energy, equivalent to what the Sun releases in about 150,000 years. Its immense power wasn't even the most startling thing about the event. As the radiation from this remote catastrophe swept past Earth, it disrupted our ionosphere and reportedly sparked mild auroral displays in the Arctic. Astronomers even detected gamma rays reflected off the Moon.

We had seen a short gamma ray burst in our own backyard, and it was deeply unsettling.

Astronomers had already come to grips with SGR 1806-20's identity as a magnetar, a neutron star so magnetized that it makes a normal pulsar look like a refrigerator magnet. Packing fields that are thousands of times stronger than a typical pulsar's and trillions of times stronger than the Sun's, magnetars are ticking time bombs. Their magnetic fields are a complicated mess, stretched and tangled and tightly coupled to the crust of the neutron star. If anything were to happen to that crust, well, let's just say the result wouldn't be something as feeble as a Carrington event.

Something *had* happened to SGR 1806-20's crust. Like Krakatoa, there had been building tension, and then . . .

Release.

On an unimaginable scale.

The instantaneous and violent starquake adjusted the crust by perhaps a centimeter or so. The neutron star, having just suffered a major stress fracture, screamed in agony, blasting out plasma and high-energy radiation. Whether a rift opened in the magnetic field or whether a complicated frenzy of events unfolded is unclear. Even less clear is whether it has the capacity for a rerun or whether this adjustment cured its magnetic tension. What is clear is that the magnetar flare was so powerful that parts of our planet felt its effects from 42,000 light-years away.

The trees didn't, though. But there's no reason that a similar, closer event wouldn't leave its mark in tree rings. So, researchers wonder, what did leave the mark? Was it the Sun? Supernovae? Magnetar flares or other neutron star phenomena? Each explanation brings its own problematic baggage.

A supernova close enough to shower us with the required energy to cause a carbon-14 spike would surely feature in ancient writings and, moreover, leave an obvious remnant for astronomers to observe. Consider that SN 1006, the most brilliant known supernova to light up Earth's sky, left no clear signature in tree rings, and you have to wonder what sort of supernova could.

Magnetar flares, though, can come from objects hiding in plain sight.

"That's one idea I particularly like," Pope conceded conspiratorially. "The energetics aren't totally wrong for getting this kind of radiation burst from a nearby magnetar. It could just sit there, minding its own business, and then every thousand years it goes off. That would be really, really cool, if it's true. Personally, I think it's less scary than some unknown kind of solar activity."

Knowing what SGR 1806-20 had done from 42,000 light-years away, I wasn't so sure about that.

Pope affirmed, "That's the hypothesis I want to be true."

My mind drifted to an *X-Files* poster I'd recently seen in an astronomer's office: *I want to believe*. As though unicorns or aliens are ever the answer.

CHAPTER 16

Cats, Rats, and Fantastic Beasts, and How to Tell Them Apart

It is not a Pandora's box that science opens; it is, rather, a treasure chest. We, humanity, can choose whether or not to take out the discoveries and use them, and for what purpose.

—JOHN SULSTON, *THE COMMON THREAD*

— A Cat in a Haystack —

On George Hobbs's computer at his CSIRO office near Sydney was something that appeared to be a colorful QR code but was actually a simulated data set. It was a square, perhaps a hundred blocks tall and another hundred long, the pixels yellow, gold, orange, or brown.

"This shows you the number of observing channels," he explained, pointing at the y-axis, which was labeled "frequency." He motioned to the x-axis and said, "This is time going across."

He clicked "play," and the not-QR-code scrolled leftward, and more pixelated columns came into the frame. The movie was not particularly interesting. Individual pixels changed from one color to another, but there was no bright dot or stripe or pattern of any kind that stood out.

"We can't find any algorithm to say that there's anything in this data set at all," he said, and for a moment, I wasn't sure what I was missing.

Then, a cat came into the frame.

Not a real cat, of course, but a cartoon cat made of bright yellow pixels in a sea of random brown, orange, gold, and other yellow ones. It was as plain as day, and I laughed out loud.

Hobbs smiled. "A human can easily see something in this data set, but because I was specifically asking for pulsar shapes, the computer is blind to this."

I was reminded of a 1999 video demonstration of something called "selective inattention." "Count how many times the players wearing white pass the basketball," read the instructions. And so, being quite sure that you are up to the challenge, you faithfully watch, making sure not to count passes between players wearing anything other than white. After the video, you are asked about the gorilla.

The gorilla? You weren't looking for one of those. But once you go back to the beginning of the video, you see the obvious. After that, it's impossible not to see it.

And there's the rub. If we want to observe things that are different from those that we are deliberately seeking, we need to be able to look for cats and gorillas. But how do you know those things are there if you've never seen them?

"Who knows what's in the actual data?" Hobbs shrugged. "That's the challenge: finding the unknown unknowns in large data sets."

So how do you find them? How many haystacks do you have to look through to find a single needle? Or a cat?

I wandered down the hall to get part of the answer from Lawrence Toomey who, as the Parkes Radio Telescope's data archives project officer, was the person best able to address that last question. In his office was a meter-tall stack of large paper sleeves, each about 40 centimeters (16 inches) on a side. In each sleeve was a plastic photographic plate from the European Southern Observatory. Thinner and lighter than the glass panes housed at the Harvard College Observatory, each plastic plate is a negative image of some patch of sky.

"They were going to just chuck them out," he said with the apologetic shrug of someone caught hoarding, "and I'd never seen these sorts of things."

So he adopted them.

Toomey described admiringly how astronomers like Henrietta Leavitt used to find transient events in the images, finally admitting, "Of course, these have all been digitized, but I just keep thinking I'm going to do something with them."

Somewhere in that stack of plates in the corner of an unassuming office in Marsfield, Australia, there might actually be something new and interesting waiting to be discovered, but even Toomey admitted he probably won't ever get around to looking through them. Instead, the pile serves mostly as a physical metaphor of a growing problem in transient astronomy. Too much data. Not enough time.

During the Harvard computers' heyday at the turn of the twentieth century, a given telescope would churn out perhaps ten photographic plates per night. Digitized, a night of observing at that telescope becomes around a hundred megabytes of data, not too much different from the photo total of a twenty-first-century teenager at lunch. Data analysis back then was done manually. Computers such as Henrietta Leavitt—not Leveret—neatly recorded sizes, brightnesses, and spectral features, plotted trends, and inferred cosmic properties. Fundamental discoveries from their analyses propelled the field forward, but even with dozens of human computers scrutinizing plates six days a week, it was overwhelming.

In modern terminology, the Harvard College Observatory's plate stacks—spanning over a century of observations—contain about ten megabytes of data per plate, give or take. There are half a million plates. The entire collection amounts to something in the neighborhood of a few terabytes of data, approximately enough to fill a generous external hard drive for a home computer.

A century later, the entire set of archived data from the Parkes Radio Telescope for the three decades beginning in 1991 amounts to a few petabytes, a word that sounds a bit like a brand of interactive toy animal. A petabyte is 1,000 terabytes, meaning that in 30 years this single radio telescope archived about a thousand times more astronomical data than the Harvard College Observatory's global network of telescopes *collected* in a century. The archiving came only after the scientific wheat had been sifted from the chaff, so to speak. Toomey estimated that the team at Parkes collected twice as much data as it kept between 1991 and 2009, and in more recent times, given the finite limits on storage, Parkes dumps far more than it can keep.

Even so, there can be plenty of surprises in those archives. Just ask Maura McLaughlin.

— RRAT Infestation —

On the door of McLaughlin's office at West Virginia University was a small poster of the original pulsar, albeit modified to have cat profiles in place of the stacked pulses made famous on the Joy Division album cover. I hesitated entering, not sure how to feel about the fact that I was wearing a T-shirt with the exact same pulsar cats, but I was soon swept into a standard academic office adorned with stacks of papers, an enormous computer monitor, and children's art. Our meeting would have to be cut short, she explained, because she was traveling that afternoon. What she didn't say was that she would be presenting the prestigious Gordon Lecture at Cornell University, an event whose previous presenters include Jocelyn Bell Burnell.

McLaughlin studies neutron stars, the havoc that they wreak on the material around them, and their usefulness as timing tools to detect any ripples in the fabric of spacetime. Her journey to studying these bizarre objects began early. "I read *A Brief History of Time* when I was a sophomore or junior in high school," she explained enthusiastically, "and I thought, 'This is the coolest thing I've ever read.' It was all about black holes and gravity and gravitational waves and spacetime."

As an undergraduate at Pennsylvania State University in the 1990s, McLaughlin was intrigued by the research of a new faculty member, Alex Wolszczan, who had discovered what had been thought impossible: planets around a pulsar. The pulsar's variable pulse had betrayed the presence of orbiting companions, which pulled the pulsar back and forth every few months, ever so slightly changing its distance and ever so slightly changing the times that we received its pulses.

This was very interesting to McLaughlin, but what really sealed the astronomical deal for her was a chance to observe some of these pulsar planets with the enormous radio telescope at the Arecibo Observatory in Puerto Rico. As she recalled her first trip there, she smiled. "It was just so, so amazing, and from that moment on, I knew I was going to study pulsars."

What set her next project apart from the mainstream pulsar research that had gone on for three decades was an interest in finding pulsars outside our Galaxy, a lofty goal for a graduate student. The tools to find nearby pulsars within just a few thousand light-years from Earth rely on their faithful timekeeping. Sure, the signal of each pulse is virtually indistinguishable from the background radio hum of the universe, but countless perfect, faint heartbeats add up to an obvious blip. For pulsars hundreds of thousands or even millions of light-years away, though, the heartbeats are too faint, and McLaughlin knew the usual tools wouldn't work. Instead, she tried to find pulsars that occasionally spat out a single stupendously bright pulse for no apparent reason. Automated digital filters could not be counted on to flag these, so McLaughlin had to return, at least in spirit, to the original method of discovery involving individual pulses. She had to look for the cats in the data herself.

Fortunately, there was already a mountain of data that had been obtained for something called the Parkes Multibeam Pulsar Survey. Parkes was where the data were obtained; "multibeam" indicated that it looked at several (13, specifically) small patches of sky simultaneously, and the term "pulsar survey" signified that this was a project that was going all-out. Astronomical surveys are never small undertakings, and entire software packages have been created to allow astronomers to sort through the data sometime before the next geological era dawns. The survey succeeded in its goal, single-handedly doubling the number of known pulsars, and it also succeeded in generating as a by-product far more data than anyone will ever fully analyze.

In this colossal haystack, McLaughlin would hunt nonrepeating radio needles.

She and her colleagues scoured four years of archived data from the survey, creating their own algorithms and flags along the way. All McLaughlin needed to do was sift through all the archived data to find a signal that was not only temporary, but also cosmic. That meant telling the computer to try out a fleet of different dispersion measures and then let her know if anything stood out from the background noise.

From the four years of archived data, she and her team found a grand total of 11 objects. Unfortunately, all of them are happily residing within the confines of the Milky Way Galaxy, each popping off intense radio waves for a handful of milliseconds at a time. They aren't the things she was seeking, but that was fine. It's not every day that a graduate student discovers a completely unknown class of object. Some of these new things repeat, but not on time scales typical of pulsars. Further data suggested that, like pulsars, these are rotating neutron stars with extreme magnetic fields. Because they are *rotating radio transients*, they were dubbed RRATs, pronounced "rats."

And the galaxy must be absolutely infested with them! Out of the 86,400 seconds in a day, a given RRAT might show up for as much as one of those seconds, and it would be observed only if the telescope happened to be pointing in the right direction. Given that the Dish was looking at patches of sky less than a square degree—which is to say less than 1% of 1% of the sky—at any given time, finding any single RRAT was incredibly improbable. To find 11 of them in the archived data meant that there must be countless RRATs, far more than traditional pulsars.

Radio astronomers wondered what else the archives were hiding. "Single pulses suddenly became interesting again," declared Simon Johnston. He added ominously, "And I've got stories about those that will curl your hair."

Keen to try a new hairstyle, I went straight to the source of the controversy.

— The Furby Craze —

Honestly, Duncan Lorimer didn't seem like the type to stir up trouble. "I've moved into administration for the time being," he said almost apologetically as he welcomed me to his office. His roles as acting chair of the West Virginia University Physics and Astronomy Department and associate dean for research would typically be enough for anyone. "But," he said. "I told them I had to keep my research going."

Lorimer had initially gotten into the millisecond pulsar game in the 1990s. "At the time, there weren't that many millisecond pulsars, and you could actually know all the names by heart," he explained, sounding somewhat nostalgic. "I was fortunate to be involved in the discovery of 0437-4715. It could only be seen by Parkes, and it was just astoundingly bright! It was so bright, we almost flagged it as interference."

Just over 500 light-years away, 0437-4715 is the closest millisecond pulsar and one of the most rapidly rotating, whipping around 174 times per second. As impressive as it is, this was not why I wanted to talk to Lorimer. I wanted to know about the field that opened up with the discovery of the Lorimer burst. It had been the first of an entirely new class of objects known as fast radio bursts (FRBs, sometimes pronounced "Furbies" by those in the business). *This* was the research that he felt compelled to continue even while wrestling with administrative duties. *This* was why Johnston had stories that would curl my hair.

To be fair, it was Lorimer's undergraduate student David Narkevic who made the discovery in 2007. Bolstered by the successful hunt for RRATs, astronomers were again sifting through the Parkes archives, but this time Lorimer was specifically hoping to find any signals out of the ordinary in the general neighborhood of the Small Magellanic Cloud. Narkevic found a singular strong radio flash hiding in the 2001 Parkes data, so strong that it had saturated the detectors for a whopping five *milli*seconds. Forget PSR J0437-4715. This radio burst was the definition of "astoundingly bright." There was no way of telling exactly how bright it was, but it was brighter than the maximum the sensors could take.

In and of itself, that would have made the burst a spectacular find. But what made it even more spectacular was just how much time elapsed between the earliest part of the signal—the high-frequency waves—and the lagging low-frequency waves. The dispersion measure was off the charts, putting the source far beyond the scant 200,000 light-years to the SMC. Far beyond millions of light-years.

Once again, it was time for the next verse of a familiar song. To saturate the detectors from such a distance, the Lorimer burst had to be immensely powerful, more powerful than anyone had imagined.

Astronomers scrambled to see if other bursts had been detected from that same location. Alas, Lorimer and his team reported, "No further bursts were seen in 90 hours of additional observations, which implies that it was a singular event such as a supernova or coalescence of relativistic objects."

No problem. The universe contains countless objects, and even though they had just one FRB under their belts, astronomers estimated that there could be literally hundreds of such events each day.

All they had to do was search tirelessly for them, preferably without stopping for lunch.

— Hunting Mythical Beasts —

Sarah Burke-Spolaor was just starting her PhD in 2007 when her thesis supervisor, Matthew Bailes, swept into her office. "He basically brought the Lorimer paper to my desk and said, 'I want you to find more of these things,'" she recalled. "So I wrote some software to find more."

Easy peasy.

Once again, it was time to scour the data, but this time for a different sort of profile. Burke-Spolaor's programs sifted through hundreds of hours of observations, flagging anything that met certain requirements. They had to be significantly above the noise threshold; they had to have particular pulse characteristics; and they had to have a dispersion measure consistent with cosmic distance.

When 16 signals passed all the tests, Burke-Spolaor and her colleagues shared an instant of triumph.

Except, there was a problem. There were 13 detectors simultaneously picking up signals from adjacent patches of sky. If something interesting were to go off in a distant galaxy, it would be visible only to the receiver beams that happened to be pointing directly at it. The Lorimer burst had triggered 3 of the 13 detectors, something not totally out of the question for a sudden bright source in space. The new bursts—whatever they were—showed up in all 13.

Nothing could be in that many places at once.

Their disappointment was palpable. The source had to be terrestrial. But why had the high-frequency waves reached the telescope before the low-frequency ones? How did the source mimic something beyond our own Galaxy while also clearly coming from Earth? And what did this mean for the Lorimer burst?

Because of their hybrid nature—seemingly cosmic, seemingly terrestrial—the 16 new signals were dubbed "perytons" after the mythical hybrid between a stag and a bird. Partly grounded, partly airborne, peryton might have been a fitting name, but it offered no solutions. Over three years had passed since the discovery of the first

FRB, and astronomers were no closer to explaining it. Worse yet, all the new evidence traced the culprit squarely back to Parkes.

— Rising above the Noise —

Operations scientist John Sarkissian held up a cockatoo feather and dropped it. As it drifted casually to the floor of the Parkes Radio Telescope control room, he began, "All the energy gathered by all the radio telescopes on planet Earth since the dawn of radio astronomy is less than the energy of that feather hitting the floor." I looked at the feather. I wasn't about to argue with someone who had been involved with the Dish since 1996, but it still seemed unfathomable to me. The telescope had six decades of nearly continuous observing under its belt, and it is just one of hundreds of radio telescopes on the planet. Sure, radio waves have low energies, but they add up, right?

Despite appearances, even the brightest objects in the radio universe aren't all that bright. Places like Parkes and the Green Bank Observatory ask you to turn off electronics when you visit because the universe is indeed whispering to us, and the radio ears we have created are incredibly sensitive.* A single cell phone on the Moon, Sarkissian told me, would be one of the strongest radio signals in the sky.

Given this sensitivity, it shouldn't be surprising that radio astronomers have to contend with countless non-astronomical sources. A passing satellite or airplane, a garage door opener, a Wi-Fi network, a cell phone, and even a robotic vacuum cleaner can interfere with radio telescope observations. Internationally, astronomy is granted certain frequency bands, but as technology marches along and as demand increases, these bands get squeezed more and more tightly. Astronomers are nearly powerless to halt the encroachment, and as a result, more radio frequency interference leaks into their observations, sometimes from unexpected sources.

With all these extraneous sources lighting up radio frequencies, it's completely forgivable that the Lorimer burst was initially met with quite a bit of skepticism. It could have been lightning, some argued, or some manmade source that hadn't been classified yet. They

* Or, more appropriately, eyes. Radio waves are light waves, after all.

would defer judgment until more were found. When it seemed like more had been found—and found wanting—many in the astronomical community turned their backs on FRB research. When a second FRB showed up in only a single beam of the Parkes Multibeam Pulsar Survey in 2012, the reception was cool at best, suspicious at worst. After all, there were plenty of radio telescopes on the planet, but only one was spotting these odd events.

A handful of new FRB discoveries at Parkes did little to allay the skepticism. McLaughlin recalled the bitterness. "We were writing proposals for grants, and people would come back and say they're not real. And we just pushed through it."

Simon Johnston echoed the frustration. "There were a few influential people who put out a paper that never got published saying that FRBs couldn't possibly be cosmological, that they were just [interference] or rubbish in the data. My view is that they set the field back five years because it was hard to get around these guys saying this."

Finally in 2014, an FRB was spotted in the archived data from the Arecibo Observatory, and the celestial nature of FRBs was almost begrudgingly acknowledged.

— A Plateful of Perytons —

That was all well and good, but it still didn't explain the Parkes perytons. There had to be a way to disentangle the mythical beasts from the true cosmic events. So, in the spirit of true scientific investigation, a PhD student and member of the Centre of Excellence for All-Sky Astrophysics (CAASTRO), Emily Petroff, along with Burke-Spolaor, Sarkissian, and others, began gathering all the clues they could, no matter how small. Was it raining on the day a peryton was observed? What time of year was it? What time of day? In which direction was the telescope pointing at the time? Anything that would help pinpoint the source.

Gradually a picture of the guilty party began to emerge. Peryton events overwhelmingly happened around midday, specifically at lunchtime, and only when the Dish was aimed in a particular direction. As eminent Canadian astrophysicist Victoria Kaspi later quipped of the whole affair, "The cosmos ought not know when it's lunchtime in Australia."

Meanwhile, SUPERB (*Su*rvey for *P*ulsars and *E*xtragalactic *R*adio *B*ursts) was ramping up, and with it came major advances in data processing and analysis. Signals from the telescope could finally be studied in real time before being shuffled to an off-site supercomputer. Now astronomers could see within seconds if a transient meeting certain criteria appeared.

The first and most obvious suspects for the perytons were the microwave ovens at the observatory. But no matter which microwave they used, and no matter in which direction they pointed the Dish, perytons remained stubbornly elusive. Until one day in early 2015, Sarkissian assumed the role of the hungry astronomer.

5...4...3...

The microwave counted down.

2...

Like someone impatient for their reheated leftovers, he popped the door open before the bell.

Three perytons appeared.

They tried different microwaves and several telescope orientations to get a more robust data set, but the answer was clear. The mythical beasts were escaping from the microwave ovens. Perytons are a cautionary tale against opening the microwave prematurely. The mechanism that generates the microwave radiation to warm your cocoa takes time—thankfully not much—to cycle down when the door opens. In that split second, it emits lower and lower frequencies until it shuts off completely. In this way, it successfully mimics the dispersion measures of objects billions of light-years away. The only thing that blew the perytons' cover was that they appeared in all 13 fields of view simultaneously.

It is admittedly a bit awkward to discover one's own microwave oven in the hunt for astrophysical phenomena. Scientifically speaking, though, the case of the pesky Parkes perytons is a success story. Not only did researchers resolve the issue of the terrestrial interference, but the explanation also lent more weight to the argument that the other FRBs are truly cosmic sources.

Tragically, the public doesn't always see the value in such a finding.

At first, the news reports were largely positive, if somewhat disappointed that the signals didn't turn out to be aliens. Petroff was invited to do a number of interviews and television spots, even appearing alongside Nobel Laureate Brian Schmidt. Not bad press for a graduate student. The frenzy then died down, and everyone assumed they could get back to work. Inexplicably, interest in the story surged weeks later, but this time with a more hostile slant.

Things got so bad for Petroff and her colleagues that CAASTRO ultimately felt compelled to issue a statement condemning the sloppy journalism, which was "damaging to the public portrayal of science" and a "gross misrepresentation of Emily's discovery and previous research." It went on to state that interview footage "was being used to effectively ridicule both Emily and Brian" and concluded emphatically that "the identification and description of perytons . . . does NOT invalidate previous research into FRBs."

And what of the research into FRBs? After the peryton kerfuffle, it gathered steam. Evan Keane, Johnston, and a growing team were on a mission to pinpoint the exact location of a Furby with an army of telescopes across every wavelength regime. In 2016, they were pretty sure they had succeeded in linking an FRB to a distant galaxy. In doing so, they also seemed to answer some long-standing questions about the material between galaxies. Startlingly, the same community that had rallied to defend the peryton team from public scorn regarded the latest announcement with scorn of its own.

Simon Johnston recalled, "It was super personal stuff. We didn't know what we were doing. We didn't know statistics. They'd say we were pulsar people, so what did we know about anything extragalactic? Then, we were chickens because we wouldn't go to conferences, chickens because we weren't on social media to defend ourselves."

Johnston gazed into the distance and added, "I haven't done anything in FRBs since."

The Lorimer burst had opened a Pandora's box. If not hair-curling, it was certainly hair-pulling and, moreover, deeply distressing. A single energetic event billions of light-years away had managed to open up rifts within the astronomical community, redirect at least one career, and strain the frequently fragile relationship between scientists and the public.

Thankfully, science news was about to be dominated by one of the most colossal events in the history of the universe, an event that came with its own song and dance. Astronomers had waited a century to witness it, and yet it was seen by precisely zero telescopes.

CHAPTER 17

Cosmic Tremors

But the Lord was not in the wind; and after the wind
an earthquake; but the Lord was not in the earthquake:
And after the earthquake a fire; but the Lord was not
in the fire: and after the fire a still small voice.
—1 KINGS 19:11–12 (KING JAMES VERSION)

— A New Kind of Observatory —

To visit the Laser Interferometer Gravitational-Wave Observatory
(LIGO) facility in Livingston, Louisiana, you need unwavering faith in
your GPS system.* Tucked in thick piney woods 30 minutes east of
Baton Rouge, this powerhouse of technological achievement—true
to the phenomena it observes—announces its Nobel Prize–winning
marvels quietly with a small roadside sign. This modesty makes perfect sense. Unlike the enormous hunting and fishing superstore I had
passed on the interstate on my way there, LIGO has little interest in
drawing traffic to its grounds.

Upon being ushered through the security gates, one begins to
sense the significance of the work carried out there. Corridors of pine
trees have been felled to make room for the enormous pipe-like tunnels that reach four kilometers to the south-southeast and the west-
southwest, a perfect right angle on the landscape. These tunnels
converge on the science facility, its stark blue and white architecture
practically an homage to right angles.

Across the parking lot is the Science Education Center, open to
anyone willing to make the trek on the third Saturday of any month.
Here, the LI (laser interferometer) part of LIGO is illustrated by dozens

---

* It's abbreviated LIGO presumably because LIGWO sounded odd.

of hands-on exhibits. Laser interferometry hinges on a well-known wave property: interference. Periodic waves consist of regularly alternating peaks and troughs, and when waves interact, occasionally the peak of one will correspond to the trough of another, canceling out the waves in that patch of space and time.

LIGO's observations start with 200 watts of laser light at a wavelength outside the range of human perception. The thin pencil of infrared light hits a beam splitter, an aptly named device that divides the light into two perpendicular beams, each of which races down a four-kilometer-long arm of LIGO. A hair over 13 microseconds later, the beam hits a hefty 40-kilogram mirror and heads back toward the beam splitter.

In a rigid universe of starched and unflappable tapestries, each beam would always travel exactly the same distance out and back. Upon returning, the beam's peaks and troughs would line up as they had initially, and the recombined beam would have the same profile as the outgoing one. But in the wobbly, dynamic universe that we inhabit, one arm of LIGO might be stretched a bit compared to its perpendicular partner. In this case, a trough from one returning wave might line up with the peak from another. The details of the newly comingled profile would show exactly how much the relative arm lengths changed and when. And that would tell scientists what caused the distortion.

In principle, LIGO is elegantly simple. Demonstrations illustrating the underlying concept have been performed on late-night television and in traveling science shows. Having seen these, you might wonder why it took a century and over a billion US dollars to make LIGO work. Moreover, why do scientists find it necessary to have so many of this sort of observatory? LIGO Livingston is tucked away in Louisiana, but 3,000 kilometers to its northwest in the state of Washington is LIGO Hanford. Across the Atlantic near Pisa, Italy, is Virgo, a right angle with three-kilometer-long arms funded by the European Gravitational Observatory. Just to Virgo's north is GEO600, a gravitational wave observatory in Germany with 600-meter-long arms. In Japan, there's the Kamioka Gravitational Wave Detector, deep underground with its three-kilometer-long arms embracing the vastly upgraded neutrino observatory that caught a dozen ghost particles from SN 1987A.

The Hanford and Livingston observatories will eventually be joined by LIGO India, a collaborative effort between the United States and India. On top of that, the next generation of gravitational wave observatories is already in the works. The coming years will see construction in the United States of the Cosmic Explorer, a LIGO-like setup with 40-kilometer-long arms, and the Einstein Telescope, a triangular observatory 10 kilometers on each side, in Europe. The Einstein Telescope and Cosmic Explorer promise to be at least ten times more sensitive than anything currently operating.

But sensitive to what?

— Pattern Matching —

Imagine someone playing a slide whistle, starting with a low note that gradually shifts to higher and higher pitches and ends with a quick, high chirp. Now imagine that the person is standing across a crowded room, and you have mild, persistent tinnitus. And there's a fountain in the corner, a ticking clock on the wall, and an exhaust fan in the adjacent kitchen. Further, the entire building is beneath the flight path of a busy airport.

If only gravitational wave detection were as easy as picking out that whistle.

Our best shot at finding ripples in spacetime lies in objects with the strongest gravity. Neutron stars will work, but black holes work even better. The problem is that physics starts to behave very oddly around things that put such steep dents into spacetime. And it gets even weirder when those things interact with each other, weirder still when they collide.

Einstein's theory of general relativity, which describes the unintuitive relationship between matter and space and time, was just the starting point for researchers like Susan Scott, a chief investigator for the Australian Research Council Centre of Excellence for Gravitational Wave Discovery. One of the things that she and her colleagues have wrestled with for decades is what kinds of ripples to expect in an encounter between two massive, compact objects.

"We hit a bit of a roadblock in the numerical relativity side of things," Scott recalled of her work in the early 2000s. The behavior of the messy, dynamic spacetime near a black hole binary could not be determined by solving a few equations and confidently circling the answer. The shape of spacetime is continually and radically changing as the system loses energy to gravitational waves and as the steep dents in spacetime warp each other. These are not easy things to figure out, even with the help of supercomputers.

She continued, "While the instruments were being built from the 1990s onwards and being made more sensitive, the binary black hole system had a number of modeling problems."

In other words, while one part of LIGO was steadily creating something capable of detecting the cat hiding in the data, another part hadn't quite worked out that they should even be looking for a cat. Perhaps a dog template was called for.

"There was a big breakthrough around 2005 by a number of groups around the world," Scott said. "That enabled us to have really accurate predictive wave forms."

You see, it isn't enough to imagine the sound of a slide whistle. Gravitational wave researchers need to know what the exact notes are. How long does it take to slide to the higher pitches? How loud is the whistle? Everything about the evolution of the frequency and loudness is crucial because it reveals something about the source. More massive binaries like black holes are analogous to louder, shorter-duration whistles. Lower-mass objects—binary neutron stars perhaps—are lingering, quieter whistles as their shallower dents in spacetime spiral more gradually toward merging into a single pit.

The 2005 breakthrough allowed scientists to model exactly what gravitational wave observatories would see if, say, a black hole with 5 times the mass of the Sun crashed into another with 20 times the mass of the Sun 2 billion light-years away, or if binary neutron stars collided 300 million light-years away. They could even disentangle information about the tilt of the binary orbits relative to Earth. Are we seeing the system edge-on? Or are we looking down at the system from some angle?

"We were ready," Scott said triumphantly. "We had the templates, and so when a signal came in, we would just measure it up against all these templates and get the one that fit."

— Needles in a Needlestack —

By 2005, LIGO Livingston and LIGO Hanford were mostly up and running after a decade of site clearing, construction, testing, upgrades, more testing, and more upgrades. And now they had the templates in hand. Or, rather, in computer. In 2006, they began the search for gravitational waves in earnest.

Five years came and went, and no signal was definitively detected. While this was disappointing, it was not ultimately surprising.

"The thing is," explained Scott, "gravity is the weakest force. And that weakness affects everything related to gravitational wave science. The [waves] sail through the material universe, and their effect on things isn't as great as you might imagine."

Scientists were looking for evidence that the arms of LIGO were stretching and contracting and that they were doing so in a way that betrayed the profile of a gravitational wave signal. But the amount of stretching and contracting expected from such an event is ridiculously small. Stupendously small. Unimaginably small. Smaller still.

Plenty of things that aren't gravitational waves can make one of the two laser beams show up a little earlier or later than the other one. Waves from the Gulf of Mexico crashing into the shoreline of coastal Louisiana or tiny seismic shifts in the state of Washington can do the trick. Then there's the force of the infrared photons hitting the 40-kilogram mirror and the fact that the structure itself is built from atoms that are not perfectly still. Any stray gas particles left in the four-kilometer-long vacuum tubes can mask the results, as can a car driving on the interstate miles away.

Gravitational wave scientists are listening for a faint slide whistle in a very noisy room indeed.*

All these complications and more mean that there are certain gravitational wave frequencies that LIGO is best suited to detect. Its

---

* There is an online game called *Black Hole Hunters* (https://blackholehunter.org /game.html) that allows players to try their hand—or ear—at picking out a chirp signal from a hiss of background static. By the time you reach the last rounds, it becomes a guessing game.

sweet spot lies between about 10 and 2,000 hertz,* but only if the gravitational waves stretch and compress the arms enough for the signal to be disentangled from the noise. For LIGO's first few observing runs, "enough" meant that the length of its arms would have to change by one part in 10 billion trillion as a gravitational wave signal swept through. That would add or subtract a mere millionth of a billionth of a millimeter from the four-kilometer-long arms. Spanning an impressive thousandth of a billionth of a millimeter in diameter, a single proton, the minuscule particle residing in atomic nuclei, is gargantuan by comparison.

And yet, by 2009 researchers had seen no hint of a ripple this significant in the fabric of spacetime. A sensitivity of one part in 10 billion trillion simply wasn't good enough to spot anything in the great cosmic demolition derby.

Astronomers had much higher hopes for the next five years. Although much of 2010–2014 was spent upgrading the observatories to Advanced LIGO rather than searching for gravitational waves, there were still terabytes of data from the previous years that promised to harbor something unambiguous. But again, there were zero confirmed detections. A string of papers on the unsuccessful searches placed plenty of limits on the types of gravitational wave sources that were detectable with LIGO, but it had clearly been a disappointing half decade.

After 2015, despite the promise of Advanced LIGO's greatly improved sensitivity, many senior astronomers remained cautious. Some had gambled on LIGO—and lost—before. Others felt that the first detections were practically inevitable once observations resumed in earnest.

"Personally, I never doubted that if the theory was correct, we would eventually detect them," Scott recalled of these uncertain days. "I knew we were going to have the sensitivity in time."

— Goliath versus Goliath —

"I see your Skype icon is green," LIGO member Eric Thrane typed to his colleague Paul Lasky. "Are you awake?"

---

* Interestingly, this is about the same frequency range as the sound waves the human ear can detect, which is why sonifying gravitational wave signals is so satisfying.

Even though it was almost 11 p.m. in Melbourne, Lasky was still up, getting ready to teach a first-year class on quantum physics the next morning.

"How excited should I be?" Lasky responded.

Very, as it turns out.

It was 14 September 2015, and the new and improved Advanced LIGO was still technically a few days from beginning its first observing run. But all the equipment was functioning as scientists and engineers took it through its final dress rehearsal. They had no idea that an entirely new window to the universe was about to be flung wide open.

Finally, the squishing and stretching of the LIGO arms was extreme enough to detect.

The ripple makers were two black holes—one weighing in at 29 times the mass of our Sun and the other packing 36 times the Sun's mass—that had spiraled toward each other in a frantic, ever-tightening, ever-accelerating orbit. As they did so, they sent off gravitational waves of increasing frequency and amplitude until they finally slammed together to create a mammoth black hole of 62 solar masses, give or take. The entire process took only a fraction of a second, and the black holes reached speeds of nearly two-thirds the speed of light in their death spiral.

The collision had happened over a billion years before the detection, which was made months before the announcement. This was nothing like spotting the obvious bright dot of a supernova. Peeling the signal from the noise required ten supercomputers to perform an analysis that would have taken a regular desktop computer 5,000 years to complete.

As is usually the case with groundbreaking scientific discoveries, the official proclamation had to wait for checking, double-checking, and triple-checking.

"I think there was still a significant amount of disbelief," Lasky recalled. "I absolutely understood the consequences if the event was real. Over the next few days, it started to sink in that this event wasn't going away, and it could not be explained any other way."

The team had to write the requisite journal article, await peer review, and ultimately, in February 2016, publish the results. During this time, the LIGO researchers effectively demonstrated that scientists should never be charged with keeping secrets. By the official end of the "embargo," the strict time frame during which results are not to be made public, everyone in the astronomical community seemed to know everything.

A professional astronomers' Facebook group had been unofficially discussing the implications for weeks, and at one point, Lasky revealed in a presentation about the discovery, a letter circulated that read in part: "Hi, all, the LIGO rumour seems real and will apparently come out in *Nature*, Feb 11. . . . spies who have seen the paper say they have seen gravitational waves from a binary black hole merger. . . . the bh masses were 36 and 29 solar masses . . . apparently the signal is spectacular."

This rumor was grossly inaccurate, joked Lasky. The journal wasn't *Nature*. It was *Physical Review Letters*.

For those familiar with the world of science, particularly regarding scientific findings that are novel or even controversial, this tale is nothing out of the ordinary. When your research team has a thousand members, there is virtually no way that any kind of secret can be kept for long. Scientists know this to be true, but the general public doesn't. That's why there are so many people willing to believe that the entirety of NASA, an organization with over 18,000 employees on its payroll, is capable of a large-scale, decades-long conspiracy to hide the "truth," whether about the Moon landings or visiting aliens or the shape of the Earth.

International news outlets exploded with the story of the discovery. LIGO scientists the world over were asked to demonstrate the slide whistle chirp that characterized the wave form. The internet brimmed with animations of enormous swells rippling through our neighborhood, each of which made it seem as if we were looking at Earth in a funhouse mirror. But GW150914 (GW for "gravitational wave" and 150914 for 14 September 2015) had done nothing that obvious. Earth itself, had it been a perfect sphere, would have been distorted by just a few trillionths of a millimeter. A still bathtub has more impressive ripples.

This isn't to say that GW150914 was a dud. During the collision, it blasted out over 5,000 foes of energy—the energy equivalent of three solar masses—as gravitational waves. In other words, in a fraction of a second, this single event expelled more energy in gravitational waves than the total emitted energy of

Every

Single

Star

In

The

Observable

Universe.

Astronomers have seen individual stars explode ten times farther away than GW150914, and those supernovae released far, far less energy but were much, much more obvious.

But this event created no light, no neutrinos, no cosmic rays. Just gravitational waves straining against the obstinacy of spacetime. Had you been only 10,000 kilometers from this cosmic showdown, you would have been stretched and squashed by a scant millimeter as the gravitational waves zipped through you at light speed.

It is a powerful testament to the cleverness of humankind that (1) we calculated what kinds of ripples in the invisible fabric of spacetime would arise from a collision like this; (2) we figured out in principle how to measure these ripples; and (3) we actually pulled it off.

"That was an amazing sensation," Lasky said animatedly, "to realize that we were witnessing and being part of history!"

"We were so lucky with that first detection. It was golden," Scott recalled fondly. "The number of firsts in that event was just astonishing, actually. First detection of gravitational waves. First direct observation of a black hole. First real confirmation that we had binary black hole systems at all in the universe."

Moreover, this discovery meant that gravitational wave astronomy had at last become an observational science. Now it was time to make it a statistical one.

CHAPTER 18

The Return of the Furbies

It sounded an excellent plan, no doubt, and very neatly
and simply arranged; the only difficulty was, that she had
not the smallest idea how to set about it.

—LEWIS CARROLL, *ALICE'S ADVENTURES IN WONDERLAND*

In March 2016, the astronomical community was still riding high on the crest of gravitational wave discovery when another major announcement was made.

"Excuse me," the object seemed to say amid the cheering and popping of champagne corks.

"Excuse me." It meekly tried again as astronomers practiced making the chirp sounds characteristic of binary black hole collisions.

"Excuse me." It persisted.

Finally, astronomers looking through archived data from the 305-meter Arecibo Radio Telescope noticed something unexpected about the fast radio burst labeled FRB 121102. It had been repeating itself.

Although there wasn't an extensive catalog of fast radio bursts at this point, this behavior was definitely out of the ordinary. Emily Petroff and her colleagues had compiled the complete list of the 16 known events, along with every measurable or inferable property, and had even optimistically created a website to keep up with new discoveries in relatively real time.*

----

* See http://www.frbcat.org, if you're keen to check.

"There were more theories than bursts," Duncan Lorimer recalled. "I mean, there were lots of variations on themes, but not a week after the original paper came out in 2007, there was already something suggesting magnetars."

That made sense. Neutron stars were an easy scapegoat. Since the discovery of the pulsar, they had shown themselves to be perfectly capable of hosting all sorts of brief, violent events, but there were as yet no widely accepted models of a fast radio burst. However, astronomers generally agreed that if something billions of light-years away rapidly emitted that much power in radio waves, it was probably the last thing that object would ever do.

When FRB 121102 managed to survive the ordeal and blast out another round of radio waves (then another, and another), Cornell University's Shami Chatterjee wrote that it "ruled out, in a single stroke, a wide range of explosive and cataclysmic theoretical models." What's more, the ten encore performances allowed astronomers to zero in on its location, the first time a Furby's home had ever been pinpointed. It was calling out to us from a dwarf galaxy 3 billion light-years away, a place where violence on this scale was completely unexpected.

As Sarah Burke-Spolaor would later succinctly state, "It was weird."

"Weird" is practically an invitation for further study, and just like that, the once maligned Furbies were suddenly all the rage. Artificial intelligence algorithms were now deployed to explore data sets on an unprecedented scale. Racing to the forefront of this data analysis frenzy was the Search for Extraterrestrial Intelligence (SETI) Research Center's Breakthrough Listen initiative, a decade-long project dedicated to finding evidence of intelligent life beyond Earth.

Created specifically to flag and categorize unusual radio signals, Breakthrough Listen's AI sifted through 400 terabytes of data from the Green Bank Telescope alone to find out what else FRB 121102 might be doing. In the process, it turned up dozens of additional bursts, 21 of which occurred within the span of a single hour. Not only was the object *not* destroying itself, but it seemed positively gleeful in its energetic outbursts.

Naturally, the public was hoping that these bursts were alien signals, especially now that SETI was involved. It didn't help matters when a highly visible astronomer suggested that the bursts were somehow the signature of alien spaceships.

Most astronomers, though, were betting otherwise. And that wasn't the only thing they were betting on.

CHAPTER 19

LIGO, We Have a Problem

Come together, right now, over me.
—BEATLES, "COME TOGETHER"

— Place Your Bets —

"Whereas," boldly began the proclamation of the public wager, "Nial Tanvir trusts his own skills in electromagnetic observations and those of his colleagues, and believes that they can overcome the challenges of large sky uncertainty regions . . ." After a few technical details, it continued:

> And whereas Ilya Mandel greatly respects the skills of electromagnetic observers like Nial, but believes that the initial impact of ground-based gravitational-wave observations will come from "dark" gravitational-wave astronomy without electromagnetic counterparts; they hereby wager that at least one (Nial) vs. none (Ilya) of the ten earliest compact binary coalescences as claimed by the LIGO-Virgo collaboration will be accompanied by a clearly identified electromagnetic counterpart observation.

This wager had been placed nearly a year before any gravitational wave observations had been made, but like Susan Scott, Mandel and Tanvir knew it was just a matter of time until those observations would be flooding in. But what would they be witnessing? Neutron star binaries had at least been independently observed, and by all accounts, neutron star mergers should be far more frequent than black hole mergers. On the other hand, because neutron stars are less massive than their black hole counterparts, the gravitational wave signals of neutron star mergers are much less obvious. During its 2015 run, Advanced LIGO was able to pick up black hole mergers from over 3 billion light-years away. A neutron star merger would have to be within just 250 million light-years to be noticeable. While 100 times the distance to the Andromeda Galaxy might seem like a long way, on a cosmic scale it's practically next door.

But the bet hinged not just on spotting gravitational waves from a neutron star merger, but also on a "clearly identified electromagnetic counterpart." For that, we would need some kind of telescope—gamma ray, visible, radio wave, or something else—to point at the direction of the source and say, "Gotcha!" If the years-long pursuit of visible counterparts for radio sources and the frustrating task of determining which galaxy just hosted a gamma ray burst had taught us anything, it was that pinpointing the source of a one-off signal is challenging. Worse, gravitational wave observatories are notoriously bad at telling astronomers exactly where to look. That's because they use interpretive dance to give directions.

— Attack of the Giant Space Bananas —

I had seen the dance performed by an assortment of gravitational wave astronomers through the years—so often that my children knew the steps. To start the gravitational wave two-step, stretch your right arm out to your side and put your left arm out in front of you, but with your left elbow slightly bent. Bring your right hand slightly toward you while simultaneously stretching out your left arm in front of you. Then bring your left arm slightly in while stretching your right arm back out. And repeat. Your right arm will be longer than your left, then shorter, then longer, then shorter. There are, of course, variations to the dance. Some dancers like to put their right arm in front of them, or even hold an arm overhead, but the idea is always the same. When one arm is stretched to its limit, its perpendicular partner is shortened.

The dance illustrates a peculiar property of gravitational waves.* As they race through space, they simultaneously alter the two dimensions that are at right angles to the direction they're traveling. If gravitational waves were coming straight out of this page, the height of the page would shorten when the width of the page lengthened. Then

---

* And gravitational wave scientists.

the height would increase as the width decreased. But if the gravitational waves were traveling from the top of the page to the bottom, the height of the page wouldn't change, while the width would alternate being narrower and wider.

Sussing out the choreography of the gravitational wave two-step from a single gravitational wave observatory is not the most precise way to determine where a signal arose, but it will at least yield a general swath of sky. What helps even more is to have a second, preferably distant, observatory whose arms have a different orientation. Now scientists can look at the two interpretive dances and combine those with information about the signal's arrival time at each observatory.

In the case of its first detection, GW150914, LIGO Livingston picked up the signal 7 milliseconds before LIGO Hanford. For light to travel directly from Livingston to Hanford takes a bit longer—a hundredth of a second—so a 7-millisecond delay meant that the gravitational waves had come from a source that was more southward than northward. Once that information was combined with the interpretive dance results, astronomers narrowed down the region for the source of GW150914 to a banana-shaped arc encompassing approximately 600 square degrees. The enormous space banana stretched across more than 1% of the sky, which doesn't seem like much until you consider that the patch would hold thousands of full Moons.

Nevertheless, ground-based and space-based telescopes immediately began scouring the sky for the barest hint that the event had produced some kind of light. *Any* kind of light. All-sky follow-ups came from the International Gamma-Ray Astrophysics Laboratory and the Fermi Gamma-Ray Burst Monitor, to no avail. The Murchison Widefield Array, the spider brigade located in Western Australia, and the nearby Australian Square Kilometre Array Pathfinder did much of the heavy lifting in the radio regime, obtaining sweeping views of almost the entire localization region. Still nothing. Meanwhile, a fleet of optical telescopes around the globe stitched together observations to create a mosaic that tiled the entire probable zone. All in all, over two dozen teams covering every possible wavelength regime tried desperately to find even a flicker of that extraordinary event, and all came up empty-handed.

Not that this was a huge surprise. Two black holes had collided, but emitting light is not really their forte even when they release 5,000 foes of energy. But some day a neutron star, perhaps even two, would be involved in a smashup, and astronomers wanted to be sure they were ready for it.

Mandel was betting that they weren't.

— Cosmic Matchmaking —

It wasn't long before LIGO had bagged its second catch: GW151012. Again, two black holes, one with 23 times the mass of the Sun and the other with nearly 14 solar masses, had collided. This collision happened over 3 billion years ago, when life on our planet was still in its infancy, and its effects rippled through our neighborhood a mere month after we opened our gravitational wave eyes. Localization for this event was even worse than it had been for the first.

Over the next year and a half, thunder from two more collisions of binary black holes rolled over us. The score was now 4–0 in favor of Mandel.

By 2017, the two LIGO observatories had begun simultaneous observing with Italy's Virgo, which provided a much-needed boost to localization. The first event observed by all three facilities was pinpointed to a much more precise patch of only 87 square degrees, but it was once again a binary black hole merger.

Five-zip.

At this point, astronomers were genuinely beginning to wonder where they had gone so drastically wrong.

"There were people who said we would never see any black hole–black hole mergers whatsoever," recalled theoretical astrophysicist Smadar Naoz. "Before LIGO made all these detections, many calculations themselves did not predict such a high merger rate."

But on the other hand, many calculations did. Predictions of the frequency of black hole mergers had always been accompanied by plenty of caveats, but given LIGO's detections, black holes had to be merging everywhere, all the time. It now has been estimated that every 15 minutes, the members of a binary black hole system crash headlong into each other and shake up the universe. This doesn't seem too problematic when you consider that virtually all massive stars begin in binary systems, but binary stars that start out relatively close to each other eventually share a common envelope. Frequently this means that the two cores merge into one, an occurrence that, Naoz points out, leaves the system with just one black hole, not two.

But what if the two stars begin life far enough apart that they each leave behind a black hole? This scenario successfully yields a binary black hole, but it creates an entirely new crisis. It would take longer than 100 times the age of the universe for gravitational waves to sap enough energy to make the two black holes merge.

"And here lies our problem," Naoz said.

Naoz has always been an astrophysicist at heart. Watching *Star Trek: The Original Series* on her family's black-and-white television in her home country of Israel, she became captivated by space at the age of five. "It opened my mind to the idea that there are other worlds," she recalled.

From that point on, she was determined to do whatever it took to become an astrophysicist, even though as a first-generation university student, she didn't quite know what that entailed. She was drawn to solving astronomical puzzles, tackling the theoretical underpinnings of everything from Jupiter-like planets in backward orbits to GRBs to cosmology. When the binary black hole puzzle came to her attention, she was interested in looking at it not from the usual stellar evolution standpoint, but from a three-body interaction standpoint. In parts of the universe that are crowded—a place like the globular cluster Terzan 5, for instance—or dominated by a single large mass, could all the gravitational jostling nudge a couple of reluctant black holes closer together?

The solution to the problem might be to look inward. Not philosophically, but literally toward the center of the Milky Way Galaxy. There, a black hole containing 4 million Suns of material lurks, just begging to throw its gravitational influence around. And the Milky Way isn't the only galaxy with a menacing heavyweight. In the 1990s, astronomers gradually came to grips with the fact that all major galaxies appear to have supermassive black holes residing in their cores.

Naoz and others modeled what happens to binary systems when they get within just a few light-years of these enormous pits in space-time, and the result was unexpected. Two objects can find themselves in a wildly chaotic orbit that is sometimes circular, sometimes highly elongated. With each close pass, the supermassive black hole can push and pull and twist the system, nudging the objects ever closer until gravitational wave radiation can finish them off.

Naoz had not intended to cause such a seismic shift in our understanding of three-body interactions when she began this work. Rather, she admitted, "I didn't understand. I wanted to redo the whole thing for myself." By starting at square one, she found that for decades, scientists had been applying the solution for one problem to a similar but different problem. "I remember showing this to friends in the meeting and with some other people there as well. And I remember I was really scared."

She hadn't been the first to question whether the effect described in a 1962 paper was being misapplied. "There were some papers, especially papers that were led by students, where there was like a foot-note. It was as if the student found something funky." She continued, "You'd think just because that's the authoritative paper on it that obviously somebody else has looked into it."

Still, Naoz would be the first to admit that LIGO's binary black hole mergers might not all necessarily be the results of countless torquing actions by supermassive black holes. Maybe there are ways that stars live and die that astronomers haven't fully modeled yet.

"I think one of the things we as a scientific community are really good at is coming up with many different solutions to one problem," she said.

The beauty of having an array of possible answers to the same question is that the possibilities frequently make different predictions. If, in fact, supermassive black holes are the matchmakers behind black hole mergers, there should be a signature in the gravitational waves that tells astronomers, "Now the orbit is mostly circular. Now it's really elongated. Now it's basically circular again."

The problem is that such a signature would have frequencies far below the ability of LIGO to detect, and the signal won't even be in the wheelhouse of the Einstein Telescope or Cosmic Explorer. Picking up such low-frequency gravitational waves will take the proposed Laser Interferometer Space Antenna (LISA), a colossal undertaking by the European Space Agency.

As its name suggests, LISA will be a space-based mission with interferometer arms spanning an incredible 2.5 *million* kilometers. Just as radio telescopes are sensitive to lower frequency light than optical telescopes, LISA's design will allow it to listen to interactions with frequencies too low for LIGO. White dwarf binaries are on its target list, as are the final stages of merging supermassive black holes. LISA will also let astronomers know that a binary system is approaching collision, alerting terrestrial gravitational wave scientists to the impending chirp. And, Naoz hopes, it will catch supermassive black holes in the act of bringing the members of a binary system closer.

"If that's observed, that will be super awesome!" Naoz said eagerly, as though the answer to her question lay just around the corner.

I looked at the calendar. "But LISA's not due for launch until—what?—the mid- to late 2030s?" I asked. "That's a long time to wait."

Naoz shrugged. "Fifteen years is not too bad."

She was right, not just in a cosmic sense but in a scientific sense. Fifteen years is less than the time it took to verify aspects of relativity or to spot neutron stars or even to localize a gamma ray burst. And for pulsar astronomers, it is just about the perfect amount of time to catch the dances of cosmic leviathans.

CHAPTER 20

Impeccable Timing

So many out-of-the-way things had happened lately,
that Alice had begun to think that very few things
indeed were really impossible.

—Lewis Carroll, *Alice's Adventures in Wonderland*

— Lost for Words —

Trying to convey the constant low background rumble of gravitational waves without mathematical terms is, as Paul Lasky discovered, almost as difficult as detecting that rumble.

"So, if we're envisioning the universe as the surface of a lake or something, and there's this flock of geese landing all around you," he attempted. "But far enough away that you're not picking up any individual goose . . ."

He trailed off.

"They should be penguins," I offered unhelpfully.

"Actually, penguins are probably better because they're below the surface, and they're coming up every now and then." He stopped and thought. "But penguins aren't in lakes, are they? I guess we'll be in the ocean. . . . But now we've got other animals and ocean waves to battle against."

He stopped again, considering his options. "I think we've probably taken the analogy a bit too far."

My son, half listening to the conversation, interjected, "And it's raining."

"Ooh, rain is actually probably better," Lasky said, brightening, and he returned metaphorically to the lake that would ripple with what scientists call "stochastic gravitational waves."

"It's raining on the surface of the lake, and I get the same ripple pattern coming off every single raindrop, but I don't resolve any individual raindrops."

I could finally see it. The raindrops—not geese, not penguins— are innumerable disturbances sloshing the lake of spacetime, and they're happening everywhere all the time. In my mind, the lake is small, the raindrops hitting quickly, but I found that I needed to expand my mind to grasp what Lasky was talking about. A single tiny splash represented the eons-long in-spiraling dance of two supermassive black holes, each containing anywhere from millions to billions of times the mass of the Sun.

We are less than a water molecule—no, we are less than a fraction of a subatomic particle—in this picture. The wavelengths and time scales of the ripples are immense compared to our tiny realm of experience.

It is a big, old, bumpy universe indeed.

Although tiny, we are ingenious. Astronomers like Lasky have realized that somewhere on that lake's surface is a way not only to detect, but also to characterize the chaos of waves perpetually rippling past. LIGO, Virgo, and other ground-based gravitational wave observatories with short kilometer-scale arms won't cut it. Those observatories are best suited to picking up only gravitational waves with frequencies ranging from about 10 to 2,000 hertz.

Even the future space-based LISA, which Naoz patiently awaits, with three arms spanning millions of kilometers will be too small, a fact that's astonishing when you consider that LISA will be able to detect gravitational wave frequencies down to about one ten-thousandth of a hertz. At such frequencies, only one wave would pass every few hours, corresponding to gravitational wavelengths of 3 trillion meters, or 3 billion kilometers. These ripples in spacetime could span the entire orbit of Saturn, and they arise from interactions that take hours to cycle. Such are the screams of unassuming compact objects, like small black holes or neutron stars, caught in the raging whirlpool near a supermassive black hole. Occasionally gravitational waves in LISA's range of vision will arise from the dizzying final circuits of two supermassive black holes.

When you get two supermassive black holes wrestling in a galactic core, though, the dance is more powerful but so . . . much . . .

slower. Forget hours. It can take *decades* for these objects to make a full orbit around each other, and the result is gravitational wave frequencies that are measured in nanohertz. That's billionths of a hertz. It would take a billion seconds—nearly 30 years—for just one of these waves traveling at the speed of light to pass you. With waves so large—100 quadrillion meters or so—you need gravitational wave observatories with arms well beyond LISA's scale of millions of kilometers. You need arms with lengths of hundreds of *trillions* of kilometers.

These are interstellar scales, and given that we haven't even managed to get a spacecraft to travel more than a few light-*hours* away, the prospects for detecting such low-frequency gravitational waves might look to be pretty bleak.

But humans are clever, and as mystifying as the universe can be sometimes, it has a propensity for handing us some surprisingly useful tools.

— Zippy's Friends Save the Day —

The key to any gravitational wave detector is knowing what to expect if a gravitational wave sweeps through and changes the distance between objects. If the distance becomes compressed, a signal arrives a bit too early. If the distance is stretched, a signal arrives late. Conveniently, nature has placed dozens of objects with extremely well-timed signals in our midst: millisecond pulsars.

The concept of using pulsars as tools to detect extremely low-frequency gravitational waves predates the discovery of the first millisecond pulsar itself, so astronomers had no idea just how stable the timing of some of these objects could be. What if, Steven Detweiler mused in a 1979 paper, pulsar astronomers turned their attention from the troublemakers of the pulsar population—the glitchers and the nullers—and instead looked at the usefulness of the peaceful, quiet, well-behaved ones? There are, he argued, a few that might be up to the challenge of becoming a cosmic gravitational wave observatory, and if that is the case, then astronomers might find, er . . .

Why, they could find . . .

Admittedly, the exact source of the gravitational waves they might detect using pulsars was still a bit fuzzy. It would have to be something with an extraordinarily long wavelength. "The close encounter of two supermassive black holes at a cosmological distance might generate such a wave," Detweiler suggested, with the caveat "but of course the existence of black holes and their interactions are matters of speculation."

The case had nevertheless been made. If there are a number of pulsars with exquisitely precise timing, astronomers could simply watch for any changes to the pulse arrival times. If pulsar A's pulses become unusually delayed and pulsar B's pulses arrive unusually early, it is possible that a gravitational wave from a powerful but lengthy process has washed over us.

The idea percolated in the back of astronomers' minds for a couple of decades. By the late 1990s, the Arecibo Radio Telescope and the Green Bank Telescope had been steadily adding millisecond pulsars to the roster, and astronomers had chosen a particularly well-behaved subset of them in a first stab at the concept. In mid-2004, the Parkes Pulsar Timing Array (PPTA) officially ramped up, using the Dish to look for any hints that gradual spacetime deformations were betrayed by the universe's most stable clocks. In 2007, the North American Nanohertz Observatory for Gravitational Waves (NANOGrav) threw its hat into the ring, and the European Pulsar Timing Array added the power of several more radio telescopes to what would ultimately become the International Pulsar Timing Array (IPTA).

Tragically, because it's located in the dynamic mess of a crowded globular cluster, Zippy did not make the cut. Teasing apart any putative gravitational wave signal from minute discrepancies in pulsar timing is hard enough without worrying about the effects of the pulsar's neighborhood. Two dozen millisecond pulsars *did* make the initial cut, a number that grew with each passing year, giving the pulsar timing arrays far more arms than LIGO. But those arms had problems that LIGO's didn't.

— You Are Here-ish —

I'm not saying that LIGO's observations don't come with their own challenges. When LIGO sees the merger of a binary black hole system, the signal carries in it information about the two masses: three numbers to describe the spin of each one (think: pitch, yaw, and roll) and how elongated or circular the orbit of each black hole is.

Lasky rattled off some other considerations. "Then there are some polarization angles, sky location, time of coalescence.... I've lost count." The computer models need to tease out well over a dozen parameters. "It's tough," he admitted. "But it's doable."

Doable, I thought, as long as you have supercomputers churning through the data.

"If I'm doing pulsar timing, though, each pulsar has lots of parameters. And that's just for one pulsar. What we're looking for are the correlations between *all* the pulsars, and the PPTA is looking at dozens."

LIGO's work was beginning to sound like child's play in comparison.

He continued, "The problem becomes absolutely intractable, so we don't do it properly. We analyze it in sort of a piecemeal way, looking at each pulsar, getting the best timing model for each pulsar, and then combining all that after the fact."

Lasky hesitated, as though preparing to tell me the really bad news. "Then, of course, there are the solar system parameters."

These are the sorts of things that most people never think about because they're irrelevant to everyday life. The reality is, though, that someone living on the equator travels the entire circumference of Earth every day as our planet whips around its axis. This means that they're moving nearly 1,700 kilometers (1,000 miles) per hour, while someone in the mid-latitudes is covering perhaps half that. The planet itself is racing around the Sun at a whopping 30 kilometers (18 miles) per *second* and not in a perfectly circular orbit. Then there's the added complication of the Moon, which acts a bit like Earth's binary companion, jostling Earth to and fro as the two waltz around the Sun. The Sun is similarly pulled around by the gravitational tugs of the planets in the solar system. Jupiter nudges it on a 12-year cycle, while Saturn's weaker tugs take nearly 30 years to play out.

As much as we've explored it, we simply don't know our own neighborhood as well as we'd like. "It's remarkable," Lasky said. "To land a spacecraft on Mars, you don't need that much precision."

But to succeed in using pulsars as very long gravitational wave observatory arms, researchers have to know exactly where all of their receivers are and what they are doing relative to the ends of those arms. Moreover, they have to understand precisely how local dents in spacetime or even coronal mass ejections from the Sun are affecting what they see.

But wait, there's more!

"We then have time-related issues," George Hobbs explained. "We measure our arrival times with an observatory clock, and we have to relate those times to the best terrestrial time standard."

He took a breath. "And *then* we have the issue that our instruments do strange things. You only need to spend some time at Parkes to realize that everything isn't perfect. Cables change. The receiver system gets upgraded. Somebody changes something."

Amazingly, the physics behind the behavior of millisecond pulsars is the least of researchers' worries. Pulsar timing arrays don't hinge on understanding the particles compressed in a neutron star's core or the exact mechanism for generating the complicated magnetic fields.

In other words, millisecond pulsars are puzzles, but at least they are useful puzzles.

— The Waiting Game —

By design, the International Pulsar Timing Array is in it for the long haul. Looking for variations over the course of decades, IPTA astronomers have no need for the sorts of instant triggers that LIGO uses to alert the electromagnetic community to a potentially observable mashup. Still, by the middle of 2015, they were hoping that our planetary ferry would have ridden over some obvious swells. With Advanced LIGO scheduled to be operational by the third quarter of the year, time was running out to be the first team to detect ripples in spacetime.

Then GW150914 crashed through.

So much for the running joke among PPTA members that showed Einstein at a chalkboard with "PPTA > LIGO" scrawled on it.

It was okay though.

No, really. It was fine.

For one thing, Lasky pointed out, the groups are complementary, not competitive, and each looks for a signal that the other can't detect. For another, there is significant overlap in the personnel involved. Lasky, like many others, is a member of both the LIGO Scientific Collaboration and the PPTA.

Scientifically, not getting an answer is, itself, an answer.

The raindrops on the metaphorical lake are countless years-long interactions between binary supermassive black holes. These interactions arise because the galaxies harboring these beasts ever so slowly and ever so messily joined forces in a cosmic mashup deep in the past. After the dust settled, what used to be two galactic hearts spiraled into one. Astronomers have countless snapshots of these galactic collisions playing out all over the universe, implying that out there somewhere, there are supermassive black hole binaries in all possible stages of the dance, each pair a single raindrop making ripples on the lake.

Our own Milky Way is not immune. In approximately 5 billion years—about the time the Sun will engulf Earth and the other planets—our Galaxy will have a life-changing run-in with the Andromeda Galaxy, spraying out stars, gas, black holes, neutron stars, white dwarfs, and more. Will our two central black holes merge quickly, at least cosmically speaking, into a single superduper massive black hole? Or is there some physical process that could prevent this from happening, sentencing them instead to circling each other for untold eons?

Astronomers don't know how things will play out once these powerhouses get within a few light-years of each other. Admittedly, they're not entirely certain about the processes leading up to the formation of these central black holes in the first place. That's one thing scientists are hoping to pin down using the pulsar timing arrays.

In 2016, after a decade of null results, NANOGrav member Justin Ellis valiantly attempted to put a positive spin on it: "We are now at a point where the nondetection of gravitational waves is actually improving our understanding of black hole binary evolution."

It's phenomenal that *not* witnessing something pushes astronomers to improve their models. Obviously, they'd prefer an unambiguous signal, but as NANOGrav's Ryan Lynch proudly pointed out, "Our secondary science is better than most people's primary science."

In the meantime, ground-based gravitational wave observatories were just warming up.

CHAPTER 21

All Together Now

We are made of stellar ash. Our origin and evolution
have been tied to distant cosmic events. The exploration
of the cosmos is a voyage of self-discovery.
—CARL SAGAN, *COSMOS*

— Turning Heads —

It was August 2017, almost exactly 50 years after neutron stars had announced themselves with a "bit of scruff" on Jocelyn Bell Burnell's printouts. For five decades, they'd been turning up right and left, but seemingly not when Nial Tanvir needed them most. Losing the wager with Ilya Mandel would mean, according to the paper they both signed regarding their bet, that he would have to organize a meeting "to celebrate the first ten ground-based gravitational-wave detections." Worse yet, he would have to begin the meeting with a speech congratulating Mandel for his victory. But there was still hope. The universe had, after all, practically handed a perfect gravitational wave signal to us on a silver platter the instant we opened the right eyes to see it. Wouldn't it be nice if it handed us another perfect signal, one involving a couple of neutron stars or even a neutron star and a black hole, when we finally had the right tools to triangulate on it?

With greatly improved sensitivity, Advanced Virgo began observations on 1 August. Across the Atlantic, Advanced LIGO was scheduled to wrap up its nine-month observing run on 25 August. The three-week overlap in operations wasn't enormous, but the universe could still come through. Moreover, the timing would be beautifully symbolic.

Yes, everyone nodded. August 2017 would be the perfect time to feel the gravitational rumblings of two colliding neutron stars.

Thankfully, the universe agreed.

With just over a week left on Advanced LIGO's observing run, astronomers caught the wave. Unlike binary black holes that rush headlong toward each other, detectable to gravitational wave observatories for only a fraction of a second, these two neutron stars lingered in a measurable dance for nearly two minutes, their weaker gravity making them more reluctant to merge. Just before their faint signal became detectable, it had a frequency of about 12 hertz. When translated to sound, that is below the threshold of human perception. Twelve times a second, two neutron stars separated by only 400 kilometers whirled around each other. Within a minute, they were circling at 30 times a second—at the threshold of LIGO's sensitivity. This frequency was finally high enough that their sonified signal became a deep bass.

A few seconds later, it was 45 hertz. By now, their separation had shrunk to just 160 kilometers.

As the gravitational waves drained energy from the stars' orbit, they grew closer and their dance became more frantic. In the last two seconds, they flew around each other hundreds of times. In these close quarters, their colossal gravity began to claw and tear at material so densely packed that it had resisted everything else the universe had ever thrown at it. Great spiraling tendrils of neutrons were ripped away and blasted into space, while what was left of the two stellar corpses joined into a single dense, seething mass. During that final fraction of a second, amid the nightmarish slaughter of two neutron stars, the gravitational wave signal crescendoed with the seemingly anticlimactic chirp of a slide whistle.

"woooooooOOOOP!" it said meekly.

And everyone turned to face it.

— Gotcha! —

Just because the universe agreed to this beautiful plan doesn't mean things went off without a hitch. For one, it wasn't immediately obvious that the gravitational wave observatories had picked up anything on 17 August 2017. The signal had sneaked up on Virgo in one of its so-called blind spots, so no alert had sounded there. Worse, LIGO Livingston had experienced the equivalent of an electronic sneeze just as the waves were finishing their final chirp, and although LIGO

Hanford had detected something obvious at 12:41:04 Greenwich Mean Time, it would take some time before scientists could pull out a tissue and clean up the Livingston data.

Less than two seconds after the LIGO Hanford alert, the orbiting Fermi Gamma-Ray Burst Monitor was hit with a short gamma ray burst. As it did with all gamma ray bursts above a certain threshold, it instantly sent out an alert that it had picked up something impressive. The short burst GRB170817A looked to be a doozy, but it wasn't immediately obvious that it was associated with anything that LIGO or Virgo had picked up.

It didn't take long for astronomers to connect the dots, and within an hour, the NASA announcement went out that there was "a preliminary identification of a GW candidate associated with the time of Fermi GBM trigger. . . . The candidate is consistent with a neutron star binary coalescence with False Alarm Rate of ~1/10,000 years."

Gamma ray burst pioneer Kevin Hurley was jubilant. "That pretty much sealed the case," he emphatically told me. "At least some of these short gamma ray bursts were, in fact, neutron star binaries that collapsed. That's the one and only example so far, but eventually we'll get a whole catalog of these short bursts accompanied by gravitational waves, and then we'll learn a whole lot about neutron stars in binary systems."

Meanwhile, the rest of the astronomical community mobilized to see what other wavelengths were telling us. But before they could do that, they needed to know where to look. Combining the directional information from the two LIGO sites yielded massive arcs on the sky, too much for telescopes to observe in any detail. Thankfully, across the ocean, Virgo's weak detection helped constrain the area. After all, there were only a handful of directions that such strong gravitational waves could come from and remain largely hidden. Once astronomers added eyewitness accounts from Fermi and from the International Gamma-Ray Astrophysics Laboratory to the mix, the patch of sky shrank even further.

Had the universe been trying to make things easy on us, the source would have been high in the sky at midnight, and the gravitational wave and gamma ray observations would have narrowed it down to a tiny patch of sky. But astronomers had to scan over 20 square degrees—enough to hold a thousand full Moons—and to make matters worse, the source was in the general direction of the Sun.

Still, with so many eyes on the sky, it took less than 12 hours for a telescope in Chile to spot a new dot in the galaxy NGC 4993, which is a mere hop, skip, and jump from here. Or, in astronomical terms, about 130 million light-years. It wasn't long before 70 telescopes on every continent fixated on that dot and tried to understand what its light was telling us.

"It was remarkable," explained Jeff Cooke, who helped coordinate many of the follow-up observations. "When we started looking at the data, we asked ourselves, 'What should we expect?' I mean, this had never been seen before, and the models were conjured up relatively recently. But when this thing happened, it was exactly what the model had predicted."

Neutron star mergers had been kicked around conceptually since the 1970s, but figuring out how much energy they could spit out and how their spectral fingerprint would evolve over time required serious computing power. Supercomputer simulations revealed a pinwheel of fireworks in just milliseconds, but beneath the flashy façade were nuclear reactions, particle creation and annihilation, neutrino-driven sandblasters, and more. Translating that to telescopic expectation was practically magic.

"Then it became a question of what to expect the next week and the next," Cooke continued. "It's just mind-boggling. . . . We've gotten to the point that we can model things that you can't even begin to describe, conditions that nobody will ever, ever encounter and that we will never even re-create in a lab."

Before long, with the help of an observational cast of thousands, GW170817 was ready to greet the public.

But first, a word from the Nobel Prize Committee.

— October Surprise —

The Nobel Prize for Physics usually lags a celebrated discovery by at least a decade. For example, Hulse and Taylor, whose 1974 discovery of the binary neutron star gradually revealed the invisible energy thief of gravitational radiation, didn't claim their prize until 1993. Chandrasekhar's 1930 calculations about the inner life of the white dwarf had to wait over half a century to be recognized by the Nobel committee. So when, just two years after the detection of GW150914, three physicists—Kip Thorne, Barry Barish, and Rainer Weiss—shared the award for their contributions toward the development of LIGO, it was a bit of a shock.

Everyone knew that it had been a game-changer, but the acknowledgment really was quick. Honestly, though, it might have been handier if the prize committee could have held off just another year. In October 2017, when the Nobel announcement was made, everyone at LIGO and beyond was busily trying to squeeze every bit of information out of GW170817, aka GRB170817A, aka DLT17ck, aka SSS17a, aka AT 2017gfo.

In a very, very closed group, it was also known as "The Event That Settled a Bet."

— The End of the Rainbow —

Mandel was happy to lose the wager. "We got very lucky," he admitted, adding, "and I had the pleasure of coauthoring half a dozen publications with my colleague Nial on the discovery and interpretation of the kilonova."

"Kilonova" was a relatively new term that burst onto the astronomical stage just seven years before GW170817. Resulting from a merger of two neutron stars or from a neutron star and a black hole, a kilonova pumps out only about 1%–10% of the light of a bona fide supernova. That's all it took, though, for GW170817 to be seen in gamma rays, X-rays, ultraviolet radiation, visible light, infrared radiation, and radio waves. But where it really shone was in gravitational waves, converting about one-twelfth of the Sun's mass into the energy required to shake the stubbornly resistant spacetime for the duration of a commercial break, or nearly a deci-foe.

In the swirling maelstrom of particles and energy, the kilonova became more than a cosmic spectacle. It also became an alchemist, forging heavier and heavier elements from the material ripped from the neutron stars. This was great news for astronomers who had been trying to sort out the cosmic origin of much of the bottom portion of the periodic chart. That explosive events were the source of things heavier than iron was practically a given.

"Exploding massive stars seemed to be the obvious site," Amanda Karakas explained. A sort of cosmic geneticist, Karakas has spent two decades piecing together how stars create and pass on various elements to the next generation. "But there wasn't any real evidence to back this up. As theoretical models got better, the idea seemed less and less likely. They just didn't produce the right initial conditions."

This didn't trouble astronomers too much, given just how difficult modeling a massive star supernova is. Maybe they just needed stronger magnetic fields, or perhaps faster rotations—or something. Still, alternative sources were suggested, and merging neutron stars had been kicked around theoretically.

Karakas said, "That was the state until the discovery of GW170817. Once electromagnetic follow-up had confirmed heavy element production, there was so much excitement. We'd finally discovered the site of the process. Yippee."

She seemed oddly unenthusiastic for someone who had finally solved a career-long puzzle. But I soon found out that even though science news outlets claimed that the process that creates all the gold and platinum in the universe had finally been caught in the act, Karakas and her colleagues were still pondering the problem. Spectra—those stellar bar codes—told them how much gold and platinum was in the Milky Way's stars, and it appeared they were too rich to be the heirs of merging neutron stars.

"All the theoretical models together still don't make as much platinum and gold as we see," she said with a hint of disappointment. "Either there's a missing source, or the models still need significant improvements."

Even so, it was estimated that GW170817 single-handedly produced as much as a few Earths of pure gold and blasted it into space, bequeathing its wealth to future generations of stars and planets.

— Looking Inward —

Gravitational wave signal.

Short gamma ray burst.

Kilonova.

The birth of new elements.

It was fine to fixate on all the amazing things that had come out of the collision, but had anybody bothered to check on what remained inside?

Susan Scott had wondered the same thing. "When I first heard the news, the first thing I said to my postdoc was, 'Oh my God, what did it form?' And nobody was talking about that."

By all estimates, the collision had resulted in a final, single object containing about 2.7 times the mass of the Sun, too massive to hold itself up against the crush of gravity.

Unless . . . could it have been spinning fast enough to withstand the crush?

The answer seems to be that, yes, it could have. At least for a while.

"I can say with a high degree of certainty that the remnant of GW170817 was so rapidly rotating that it was either a hypermassive neutron star or a supramassive neutron star," Paul Lasky stated.

These are quite possibly the weirdest objects that both the universe and theorists have ever conjured up, the difference between "hyper" and "supra" ultimately hinging on what exactly goes on inside the neutron star. In either case, the magnetar spins so fast that the centrifugal forces combined with the pressure of neutrons are able to hold off the gravitational collapse just . . . a bit . . . longer. Conspiring with gravity is its overzealous magnetic field, which puts on the brakes. Eventually, even a hypermassive neutron star will slow to the point that it gives in to the inevitable.

"It definitely didn't form a black hole instantly," Lasky assured me. But how long did it last before it caved in?

The exact timing isn't certain. The remnant object from GW170817 might have enjoyed half a second or even a leisurely half hour before its rotation slowed enough for gravity to finally break down its material barriers and plunge the entire thing into an infinitely deep well in spacetime.

But in the end, gravity did win.

FADE TO BLACK.

CHAPTER 22

Multiple Eyewitness Accounts

Alice watched the White Rabbit as he fumbled over the list, feeling
very curious to see what the next witness would be like, "for they
haven't got much evidence *yet*," she said to herself.

—Lewis Carroll, *Alice's Adventures in Wonderland*

— Ostentatious —

It might have been the end for the two neutron stars whose dying
screams resulted in GW170817, but it was a new beginning for multi-
messenger astronomy. For the first time in human history, astronomers had measured not just the light from a cosmic event, but also
the gravitational waves.

Of course, astronomers had gotten two stories from different cosmic envoys before. SN1987A yielded both light and the neutrinos that
astronomers expected from a core-collapse supernova event. And for
decades, astronomers have picked up cosmic rays, energetic subatomic
particles that race through space, during solar flares.

But GW170817 was different. It wasn't some event that was practically next door. It was *out there*, and it gave us everything we could
have hoped for. Gravitational wave astronomers witnessed via our orbiting observatories the shrinking spiral of two neutron stars that collided
and blasted out a two-second burst of gamma rays. For decades, researchers had hoped for such an obvious signal that would pinpoint
not just the location, but also the mechanism behind a GRB, and the
universe delivered. Then there was the lingering glow of the kilonova
across the entire electromagnetic palette, an observational gold mine
for astronomers to pick at for years to come. The collision had so
many ways to tell us about itself that it almost seemed like it was trying

too hard to be noticed, particularly for something that generated less than a tenth of a foe of energy.

Meanwhile, a supermassive black hole deep in the universe was about to see what astronomers could do with a single neutrino.

— The Big Chill —

One neutrino.

That's all it took.

One unfathomably tiny, energetic subatomic particle made a 5.7-billion-light-year trek, unperturbed by the intergalactic and interstellar material that stops so many other messengers dead in their tracks. When it finally got to Earth in September 2017, its immense energy—more than 20 times what the Large Hadron Collider can impart to any particle—triggered a cascade of events deep in the Antarctic ice.

Thankfully, the aptly named IceCube was waiting for it.

The IceCube Neutrino Observatory is astronomers' latest entry for the title of Most Unusual Observatory, whose reigning champion for years had been tree rings. Even when it was being constructed, IceCube rapidly shot to the top of the rankings. As its name suggests, IceCube is an observatory shaped like a cube. In the ice. Buried 1.5 kilometers beneath the South Pole, to be exact. It contains thousands of optical sensors strung in an enormous latticework spanning one kilometer in every direction. It took a 500-ton drill and a fleet of scientists, technicians, and engineers seven Antarctic summers—the only time it was warm enough and light enough to work—to drill the holes using jets of boiling hot water. After that, they strung dozens of cables with thousands of basketball-sized sensors and slid them down the tubes like extreme ice fishers. Finally, in late 2010, after a slow but steady start, the observatory was ready to catch some of the universe's fastest and most energetic particles.

IceCube doesn't actually see those particles. Instead, it detects the tracks they leave behind in the ice. Specifically, it picks up the eerie blue glow of the Cherenkov effect as a high-energy neutrino slams headlong into an atomic nucleus in the pristine, transparent ice. The process creates a high-energy muon that races away at speeds exceeding the speed of light, creating a "luminous boom" in its wake.

On 22 September 2017, just a month after the astronomical community heard about a neutron star collision from two independent messengers, another message from another catastrophe came our way. A supermassive black hole had (possibly) just ripped apart a star.

— Nice Shootin', Tex —

The bullet it fired at us was a neutrino. This was no ordinary neutrino though. It packed literally millions the amount of energy that a typical supernova neutrino carries, meaning that whatever fired this particular shot possessed energies that would make a supernova envious.

As the resulting muon raced through IceCube's thousands of detectors, it left the luminous equivalent of bread crumbs that pointed back home. Astronomers quickly identified the gun as belonging to TXS 0506 + 056, a bright radio and gamma ray emitter that I will henceforth call Tex.

The neutrino alert was a wanted poster sent to anyone who might have witnessed anything unusual. "Have you seen this monster?" it asked. The Fermi Large Area Telescope, part of the Fermi Gamma-Ray Space Telescope, watched the entire gamma ray sky 24/7. Surely, it saw something suspicious. What about MAGIC? The 17-meter telescopes comprising the Major Atmospheric Gamma Imaging Cherenkov observatory in the Canary Islands might have spotted their own Cherenkov glows in the sky around the same time. The call went to Namibia to see if the High-Energy Stereoscopic System had seen Tex, and to the Very Energetic Radiation Imaging Telescope Array System in Arizona. Radio observatories were called into action. The 27 dishes of the Karl G. Jansky Very Large Array checked for anything unusual, as did the 40-meter radio telescope at the Owens Valley Radio Observatory. Astronomers even sent an assassin after Tex. The All-Sky Automated Survey for Supernovae (ASAS-SN, pronounced "assassin") was employed to see if Tex had done anything visibly unusual.

And these were just a few of the observatories and telescopes that rallied to IceCube's side.

Soon, astronomers were satisfied that they had enough evidence to implicate Tex for firing the shot. At the time the astonishingly energetic neutrino came blasting into our neighborhood, Tex was also unusually active in high-energy gamma rays.

Plenty of things flare in gamma rays without blasting high-energy neutrino bullets at us though. Tex is always a fairly bright source of both gamma rays and radio waves. What sets Tex apart is that it's 5.7 billion light-years away, the barrel of its gun a cosmic jet blasting material away from a 300-million-solar-mass black hole that is busily shredding everything around it.

Tex is a blazar.

Astronomers have known about Tex since the 1980s, but it took nearly two decades to become fully acquainted with its immense power. It's in a category of objects known as "active galactic nuclei," and the name basically says it all. They are the centers of galaxies, and they are terrifyingly, violently active. Each is powered by a central supermassive black hole, but unlike the relatively tame neighborhood around the Milky Way's supermassive black hole, an active galactic nucleus contains a cosmic whirlpool of infalling material. As it races around the central supermassive black hole, the scorching hot matter gives rise to intense magnetic fields that squeeze jets of plasma from the poles. The jets shoot out at nearly the speed of light, a warning beacon to the rest of the universe about the monster beneath.

The brightest active galactic nuclei can be seen almost to the ends of the observable universe. They are the quasars that mystified astronomers in the 1960s, too energetic to be galaxies and too redshifted to be stars. The most luminous quasar is home not just to a supermassive black hole, but to an ultramassive black hole, weighing in at 12 billion times the mass of the Sun. As this monster shreds everything that comes near it, firing out beams of light and particles like titanic blowtorches, it glows with the energy of 400 trillion Suns. That's over 100 foes per day, every day, until the whirlpool feeding it runs dry.

Sometimes, if we're lucky, observationally speaking, we get a chance to peer down the barrel of one of those blowtorches and see a blazar like Tex. It's no danger to us. Blazars are all necessarily a long way away, with the closest one being fully 2.5 billion light-years distant. As galaxies age, they settle down quite a bit, like a puppy that finally stops gnawing on everything. But until then, they shred the living room furniture with unbridled enthusiasm.

The blowtorches of an active galactic nucleus are like a perpetual particle accelerator. Protons crash into neutrons. Neutrons sideswipe electrons. Light spontaneously morphs into matter, which might morph back into light. Particles and antiparticles and photons slam into each other with reckless abandon, and if the supermassive black hole decides to eat a particularly big meal one day, that just makes things more exciting in the supercollider.

Every so often after a big lunch—possibly a star that wandered just a bit too close to be able to hold itself together—the blazar might flare up in gamma rays. During that bout of galactic indigestion known as a "tidal disruption event," a couple of relativistic particles in the jet might crash headlong into each other, creating a super-energetic neutrino that races unhindered through the universe, never altering its course. If that happens in a blazar like Tex, that neutrino might even head straight toward Earth, where over 5 billion years later, clever creatures with eyes in the ice, on the ground, and in the sky might even be able to trace it back to its origin.

— Observational Overload —

Tex is one more example of the urgency of time-domain astronomy. Gone are the days of leisurely sending out a telegram telling other astronomers to take a look at a new dot in the sky when they get the chance. Modern transient astronomy demands action—and demands it now.

Jeff Cooke knows the pressure all too well. He was drawn to instant astronomy through, of all things, fast radio bursts. Hoping that FRBs would be accompanied by some other type of light—*any* other type—radio astronomers had taken a stab at coordinating with other telescopes to solve the riddle. Their efforts had come to naught. Then, Cooke offered a solution.

> The problem is that with a millisecond burst, once the radio telescope says, "Hey, we found one," it's already too late. No matter

how fast you are, you've already missed it. So I set up a system to coordinate optical, radio, UV, X-ray, and gamma ray [telescopes] to be observing concurrently with the radio telescope, and when they find an FRB, we'll already be right there observing.

From this nascent effort in the early 2010s grew the Deeper Wider Faster (DWF) program, which now coordinates dozens of telescopes across every wavelength regime to catch astronomical transients in the act.

Cooke painted a picture of the process. "When it's running, there can be as many as 30 or so people in the room, and it's like the floor of the stock exchange. We're trying to analyze data that is coming in *right now*, that's happening on the sky *right now*. Should we trigger on this?"

And that's a tough call. It's not as though every astronomer on the planet is concerned with transient events, and other researchers' observing projects can't be pushed out of the way just because some other telescope picked up a new flash. It's pretty bold to tell someone who's working on a world-class telescope to drop what they're doing *right now* and slew over to another patch of sky.

"Things can get a little tricky," DWF's Sara Webb explained. "An extragalactic transient is hard to spot because it can look like something that can be attributed to plain old stars living their lives."

The fear of interrupting someone's observing session isn't the only reason to be cautious. It turns out a mistake can also be very expensive.

"We've almost followed up [false alarms] with one of the big telescopes that cost two dollars per *second* of observing time, so you want to be able to identify those normal events because they are wasting your time and money," she explained.

Webb and others around the globe spend their time not just trying to understand the signatures of the truly significant transients, but also telling computers how to spot everything the universe throws our way. "You're trusting the computer to identify patterns," she said. "And we're starting to realize that computers can be fantastic at identifying patterns."

That's a good thing, particularly with the raging torrent of data that astronomers are now producing. But there is a downside.

"Basically, you're trusting everything to a data broker, trusting it to evaluate your data without physically looking at it. Then it outputs what it thinks you want," she said.

Cooke's analogy to the stock market floor had been spot-on. Many astronomers now work with alert brokers, makers of computer algorithms that can instantly characterize a signal and decide if it's something that researchers might want to follow up on. This approach is particularly important in this era of large, data-intensive surveys.

In the spirit of Fritz Zwicky's early surveys, the Zwicky Transient Facility in California uses a wide-field imager to sweep the entire sky in just two days, spying on literally billions of objects. This survey generates terabytes of data per day and alerts astronomers to millions of potentially interesting occurrences, most of which are actually uninteresting to most astronomers most of the time.

But this is nothing compared to what the Vera Rubin Observatory, formerly known as the Large Synoptic Survey Telescope, will produce when it's completed.

"That is going to be one of the biggest surveys ever," Webb said with a mixture of excitement and trepidation. "The amount of data coming off it is going to be impossible. Impossible. Even if you get every single astronomer in the world looking at it."

With a massive field of view of 9.6 square degrees and targeting literally tens of billions of stars and galaxies, the Rubin Observatory stands to generate a petabyte per week.

A petabyte. Per. Week.

In that firehose of data, there will be 10 million transients flagged every single night. As a result, astronomers will find themselves simultaneously exploring uncharted waters and paddling around their familiar lagoons.

"We need to decide where we want to go with astronomy as a whole," Webb said. "In transient astronomy, will we be limiting ourselves to what we know right now? I mean, we can build a computer that's looking for something that you've modeled, like a kilonova, and you might have built the model with the best knowledge at the time, but every now and again, the models will need to be checked and compared to different data."

She added with a touch of nostalgia, "The next generation of astronomers might not ever even observe with the telescope because computers are doing it all for you."

On the other hand, this is a situation that radio astronomers have experienced for decades. Just because astronomers aren't sitting at a telescope on a remote mountaintop doesn't mean there's no connection with what they're observing.

CHAPTER 23

Furbies—A New Hope

> "Have you guessed the riddle yet?" the Hatter said,
> turning to Alice again.
> "No, I give it up," Alice replied: "what's the answer?"
> "I haven't the slightest idea," said the Hatter.
> —Lewis Carroll, *Alice's Adventures in Wonderland*

— Chiming In —

"What is that?" Emmanuel Fonseca remembered asking himself as he stared at the giant glowing V in the sky. "I couldn't believe just how wondrous it was." It was 1997, and this was the scouting trip that would seal his future. Sure, he had been curious about space before, but the spectacular appearance of Comet Hale-Bopp tipped the scales. He wanted to become an astronomer.

"I'm the only one in my family to pursue science as a profession," Fonseca said, hastening to add, "and they're all really supportive and caring about it. But it's very much a sort of black sheep thing where I'm from."

By his own self-deprecating account, he stumbled through high school, university, and graduate school, latching onto pulsars along the way: "I find them so intriguing and useful." He continued enthusiastically, "You can make all these astounding, groundbreaking measurements just by studying a certain object and using its timing properties."

From there, he went down the road paved by so many pulsar aficionados: FRBs. "In 2016, I began helping to build a telescope in Canada, the Canadian Hydrogen Intensity Mapping Experiment [CHIME], literally creating maps of the sky in radio frequencies," he said.

Hydrogen intensity mapping seems to be a long way from pulsars and FRBs, but the unusual telescope CHIME has plenty of tricks up its sleeve. Instead of using a dish that aims at a particular location in the sky, CHIME uses what looks like four enormous parallel half-pipes in a field at the Dominion Radio Astrophysical Observatory, a few hundred kilometers east of Vancouver, British Columbia. Its unusual design allows it to see 200 square degrees of sky at any given time, and as Earth rotates, more objects sweep over the telescope.

Even though the installation's stated mission was mapping hydrogen intensity (not Canadian hydrogen, but cosmic), astronomers quickly realized that CHIME had an uncanny talent for spotting temporary radio signals, like pulsars and FRBs.

"You know," Fonseca said. "Things that go bump in the night."

Constructing CHIME was very much a grassroots effort, with students, postdoctoral researchers, and faculty members painting everything white (to reflect sunlight), pulling cables, and installing hardware. With the exception of the main steel structure, the entire telescope was built by the people who would be using it. This approach was reminiscent of Bell Burnell, who had strung kilometers of wire for her pulsar-discovering telescope 50 years before. And like Bell Burnell, Fonseca has a deep connection with the instrument he uses.

"When you're using [the] Hubble Space Telescope, you use the data it sends you. You didn't have a hand building or designing the telescope," he explained. "I think people saw the promise in the concept of CHIME for a long time, but for these things to exist, you need a dedicated team of people willing to put aside their ability to do science in the moment to enable the science later."

In the grand scheme of astronomical observatories, "later" wasn't a particularly long time. The concept for the telescope arose in the early 2000s after the accelerating expansion of the universe threw everyone for a loop. By 2013, construction was under way. Fonseca joined in a couple of years later, and in early 2018, CHIME was ready for action.

Continuously sweeping the sky on a daily basis, while a surefire way to pick up plenty of transient events, is obviously a data-intensive endeavor.

"Netflix levels of data," Fonseca stated.

But what they're interested in isn't a database of movies. They're just hoping to find each instance of, say, the word "unicorn," and so the team has created efficient screening algorithms that dump most of the data before it even gets off the observatory site. After all is said and done, the total data exported comprise only a few gigabytes per day.

Only.

CHIME made its first FRB detection almost instantly. Then it found eight new repeaters. If astronomers were ever going to get a handle on where these things live or what causes them, the answers would most likely stem from tracking these repeat offenders.

Nobody guessed that we might get a call from inside our own house.

"Inside our house" is, of course, still 30,000 light-years away, but it is in the Milky Way Galaxy. The surprise Furby came from a soft gamma repeater known as SGR 1935 + 2154, which lies in the general direction of the Galactic center. Observing such a powerful pulse of radio waves from a known magnetar was unprecedented but not completely unexpected, and it seemed to answer at least part of the question.

However, as the saying goes, a zebra might look a lot like a horse, but it's not a horse. And while FRBs look a lot like magnetars, it's a leap to say they really are the same species. Yes, *some* FRBs can be produced by a flaring magnetar, but what popped off in our house was far too weak to explain the things that had dispersion measures putting them billions of light-years away.

Thank goodness, really. We already know what a powerful magnetar flare can do to us.

"The elephant in the room is that we don't really even know how pulsars work, fundamentally, and with FRBs it's even worse," Fonseca admitted. "At least we know pulsars are spinning, magnetic neutron stars. We just don't really know what FRBs are to begin with."

One nice thing about FRBs—like so many other energetic events in the cosmos—is that we don't have to understand them for them to be useful.

— Furbies Unite —

Simon Johnston had hinted at FRBs' usefulness, but it seemed quite a stretch for something so elusive and so mysterious to be good for anything. By the time I talked to him, only a dozen or so had been pinned to a host galaxy.

Then again, the field had barely taken off its training wheels.

"At the moment, only about 4% of the universe is baryons. You know, the stuff that we're made out of," he explained. "But of that 4%, we only see about half. And I'm not talking about the dark matter."

Ah, yes. Dark matter is another matter entirely. Making up about 27% of the universe, while the energy that drove the unexpected accelerating expansion of the universe accounted for the rest of the universal inventory, dark matter has been precisely measured and modeled, but never seen. Not with any kind of electromagnetic radiation, neutrinos, or even gravitational waves. But that isn't the missing stuff that Johnston was discussing. We had only ever succeeded in seeing half of the normal, everyday stuff that we knew *had* to be there.

"Presumably, the other half was in the very tenuous spaces between galaxies, which makes it very difficult to see with any instrument," Johnston continued.

"But," he declared with a victorious stab at the air, "the dispersion measure tells me exactly how many of these baryons are between us and the radio source, so when we measure that number and we know the redshifts of the host galaxies, then you can directly measure the number of baryons!"

He paused to let me fully grasp the implications. Fast radio bursts yield radio waves. Higher frequencies get here first. How much faster they arrive gives astronomers the dispersion measure. The dispersion measure tells us the amount of stuff that the radio waves played with along the way. The redshift tells us the amount of space. Which means . . .

Putting them together tells us how the stuff is distributed in space, even when we can't directly see that stuff. And this is something that nobody had ever succeeded in doing before.

The work Johnston was describing had been spearheaded by Jean-Pierre Macquart, who died suddenly and unexpectedly in mid-2020 just after announcing that half of the universe's material had finally been found. The relationship that he discovered between redshift and the dispersion measure is now called the "Macquart relation" in his honor.

Johnston looked to the future. "You can do serious cosmology, really test the models that we've got. We're talking the Hubble constant and all that. That's why people are very excited. They're not excited about FRBs because there's some weird thing that goes BANG! They're excited about them because you can do cosmology."

But they won't be able to do cosmology with the scant fraction of repeating sources. "If we want to do this properly," he said emphatically, "we've got to get very good precision for the one-off events."

A hundred would be a good start. Ten thousand would be much better. That means astronomers need to do for FRBs what they did for GRBs: find a bajillion of them and localize instantly.

CHIME is trying to do its part. In 2021, it dumped 500 new FRBs on the laps of astronomers, single-handedly expanding Petroff's once optimistic online catalog by a factor of six. Every day, CHIME churns out a few more, but unfortunately without precise enough localization to pinpoint their host galaxies.

Yet.

I had seen one of CHIME's little siblings under construction at the Green Bank Observatory. It was just a few piers in the concrete at the time. This and two other "outriggers" (in California and British Columbia) promise to help astronomers focus on the sources because they act like a single radio telescope the size of North America. Then there's the Australian Square Kilometre Array Pathfinder, a vast array of 36 radio telescopes near the Murchison Widefield Array in Western Australia, China's Five-Hundred-Meter Aperture Spherical Telescope, and the Y-shaped Karl G. Jansky Very Large Array in New Mexico, just to name a few of the heavy hitters.

Admittedly, this isn't quite the way CHIME scientists had envisioned their quest for H_0 and q_0. CHIME had originally been conceived to hunt down the very first bumps in the cosmos, bumps that occurred long before the first massive stars exploded, long before the first fast radio burst or star-eating quasar. Indeed, these bumps happened long before the universe had even managed to pull together atoms, much less stars.

You might have heard it said that in space, no one can hear you scream. But at the time of these first bumps in the universe, there was nothing but screams.

CHAPTER 24

The First Bumps in the Universe

Plus ça change, plus c'est la même chose.
[The more things change, the more they remain the same.]
—Jean-Baptiste Alphonse Karr, *Les Guêpes*

— First Impressions —

"Let's go right back to the beginning of the universe," Tamara Davis began, and then she took me back to a time when the entire universe— *the entire universe*—was denser than the unfathomably dense interior of a neutron star and hotter than the hottest stellar interior.

Somewhere between birth and a hundred millionth of a trillionth of a trillionth of a second later (0.0000000000000000000000000000000001, or 10^{-32}, seconds), our universe ballooned to trillions upon trillions of times its original size, stretching the fabric of spacetime so tightly that random variations that existed before then were almost completely smoothed over.

Almost.

Some pockets of the universe ended up with densities that were ever so slightly greater than their surroundings, creating the shallowest of valleys in spacetime. If the universe were a trampoline one meter off the ground, these early depressions would have been less than a millimeter deep. There was no rhyme or reason to them, just as there is no rhyme or reason to the popularity of some internet memes over others, but their presence created the scaffolding of the material universe.

Into those initial shallow depressions flowed gravitating matter. Mixed into this dense cauldron of electrons, protons, and neutrons— the latter two collectively known as baryons—was light, bouncing from charged particle to charged particle and pushing at them as it tried to find some escape from this claustrophobic nightmare. As the matter jostled toward the centers of the gravitational valleys like eager concertgoers toward a stage, the photons pushed back and created compression waves that swept outward from the dips at the speed of sound. But these were not sound waves in air, a cold, empty medium that propels the waves along at a snail's pace of 343 meters per second. This was an unimaginably hot and dense fluid of photons and particles, a place where acoustic waves could travel at upward of 170 million meters per second, or nearly 60% of the speed of light, far faster than the supersonic shock wave racing out of the interior of an exploding star.

Which brings us back to the tiny variations in the inflating universal balloon. Complicating the early situation was the presence of dark matter, a substance that makes up over 80% of the gravitating matter in the universe, but that astronomers have yet to put their theoretical fingers on. What we do know is that dark matter doesn't interact with light, so while the baryons and photons were pushing each other around, dark matter continued sliding unobtrusively into the small pits in spacetime, making those pits bigger. Once ensconced, the dark matter tried to coax the massive baryons toward it, but the light continued to fight back. Wave after wave of baryons fell inward, only to be pushed outward by the multidimensional snowplow of photons.

Meanwhile, the universe continued to expand. Baryons continued trying to fall into the dark matter clumps, and light snowplowed them away. Around the 380,000-year mark, something new happened. The conditions in the rapidly expanding and cooling universe allowed the first atoms to form as electrons and nuclei came together in an event known as "recombination," although most astronomers agree that it should simply be "combination." Because neutral atoms are largely immune to light's push, the photons were finally released from captivity and for the most part have never been bothered by matter since. In fact, this freed light has stretched with the expanding universe and now fills up the entire sky. It is known as the "cosmic microwave background," one of the most compelling pieces of evidence for the big bang theory. Without the perpetual push of the snowplows, a pileup of snow (read: baryons) was left at a very specific distance away from that central dark matter concentration. It is called the "sound horizon."

Since the late 1960s, cosmologists have known there should be a greater concentration of matter at the last point of contact between photons and baryons, just as there should be a central peak where the dark matter set up shop in the shallow dents. In some respects, it would look very much like the snapshot of a pond just after a pebble was thrown in.* There would be a central uplift and a ring whose radius was determined by the time the wave had traveled and the wave's speed, which was dependent on the conditions of the fluid.

In a convenient universe, there would be only a single pebble dropped into the pond. In the real universe, though, things are much messier.

"It's like a whole bunch of pebbles were thrown at all different times into the pond," Davis explained, pulling up an animation to illustrate the point. "As a result, you get all these overlapping patterns out of it." The outcome was a cacophony of interfering waves whose patterns became imprinted on both the material universe and the distribution of light itself.

"It's actually one of the most profound things that I've learned about the universe," Davis said. "Even if you start from completely random initial conditions, you can emerge with structure, and that structure won't be random."

I stared at the last frame of the animation. The peaks of the choppy water had morphed into luminescent clusters of galaxies, and the dips had become voids. Try as I might, I could not see anything resembling a pattern in those galaxies. This is another place where astronomers need tools that can spot the structures that the brain can't.

— Galaxies Galore —

Davis pointed at a single dot representing a single galaxy sitting atop the crest of a ripple. "Basically, we start with a ruler on a particular galaxy and measure the distance to every neighboring galaxy."

"Oh, is that all?" I thought, as I considered the number of dots representing galaxies on the screen.

---

* Tossing things into ponds is, I've discovered, fundamental to explaining some of the more esoteric astronomical concepts.

"Then we count how many galaxies are separated by each different distance," she continued, and she hit a few more keys.

On the screen, an animated ruler popped up. If the measured distance between galaxies was, say, 50 million light-years, a vote was added to the corresponding column. A bar graph with the vote tallies slowly constructed itself. At first, it appeared somewhat random, with equal numbers of widely separated and closely snuggling galaxies showing up. Even after the animated ruler had made thousands of virtual measurements, there was nothing obvious.

Davis explained the holdup: "It's less than 1% deviation from homogeneity. That's why we could only see it once we had . . . hundreds of thousands of galaxies."

One of the first teams to attempt spotting the baryon acoustic signature was the Australian Two-Degree Field Galaxy Redshift Survey, begun in the late 1990s. It made a tantalizing, if ambiguous, "possible detection," and was followed up on the other side of the Pacific by the Sloan Digital Sky Survey, which kicked into action in 2000. Using its dedicated 2.5-meter telescope, researchers carried out a small number of large-scale projects that ultimately required thousands of hours of telescope time spanning years and capturing huge swaths of sky. In 2005, after looking for correlations in the separations between pairs of 46,748 galaxies, astronomers announced that galaxies could indeed be found on cosmic ripples of a particular size.

A similar survey called WiggleZ ran from 2006 through 2011, getting positions and redshifts for a quarter of a million galaxies. Its goal was to use baryon oscillations (hence the "wiggle") at vast distances (higher redshifts, aka higher Z), looking back at a time when the ruler showed that the universe was a bit smaller than it is today. And, yes, WiggleZ was a nod to the popular Australian children's entertainers the Wiggles.

This work clearly wasn't happening one galaxy at a time. At one telescope was an elegant system with 400 optical fibers and a robotic mechanism that placed the fibers in such a way that the light from a single point in the sky entered each one. With 50 fibers devoted to calibrating objects and 350 that captured light from target galaxies, the team was able to observe the spectra of hundreds of galaxies per hour, yielding thousands on a good night. For the Baryon Oscillation Spectroscopic Survey, part of the next generation of the Sloan Digital Sky Survey, which ran through 2014, there were 1,000 fibers. When I talked to her, Davis was working with the Dark Energy Spectroscopic Instrument (DESI), which can get 5,000 at a time.

"It took three days for DESI to get what we did in five years with WiggleZ," Davis said. "It's quite depressing, really."

Observing that many galaxies comes with an added bonus.

Davis pulled up another animated data visualization: a 3-D rendering of the locations of the galaxies studied by the various projects over the years. It looked like an enormous sparkling fluff of dandelion, each hair a line of sight that encompasses galaxies spanning billions of light-years. The whole thing glistened like a shower of glitter.

"These lines are the regions of sky that we looked at with something called the Dark Energy Survey,* and the sparkles are all the supernovae that went off," she said nonchalantly. "I sometimes give talks to amateur astronomy clubs, and a few of those people have discovered supernovae. Someone had discovered two, and that's definitely impressive, but then they asked me, 'How many have you discovered?' and I was like . . ."

I regarded the shimmering glitter, waiting to hear the answer.

"A few thousand."

I blinked in disbelief.

She added hastily, "But to be fair, I was using professional telescopes and looking at hundreds of thousands of galaxies. Honestly, I think my achievement is . . . less impressive than what they've managed to do with the instrumentation they have."

Also to be fair, Davis wasn't acting alone. At the time, she was part of a team of over 400 researchers from institutions all around the world.

Out of those thousands of supernovae, Davis estimates that there are good data on about 1,500. "Those are the really high-quality ones that we can use for cosmology."

---

* Abbreviated DES, this project shouldn't be confused with DESI but frequently is.

I thought about Fritz Zwicky, whose stubborn determination yielded more than 100 supernovae over his lifetime. The High-Z Supernova team and others had caught only a few dozen high-quality ones in the 1990s, a figure that was enough to turn our idea of the expansion of the universe on its head. Davis was talking about 1,500.

She turned again to the graph showing distances between galaxies. "On the large scales, it peaks at about 150 megaparsecs," Davis said. That means that there's an unusually large number of galaxies sitting 500 million light-years away from each other.

To guide my eye, she overlaid the individual rings from each sound wave snowplow on top of the galaxies, and the pattern was finally visible. It was a mess, but it was visible.

— Setting the Standard —

The DESI team will eventually make a precise three-dimensional map of 30 million galaxies, reaching almost to the ends of the observable universe. This information can be combined with data from the Dark Energy Survey, which has captured half a *billion* galaxies. But making a map is not the end goal.

"These baryon acoustic oscillations are things of known size," Davis explained. "So they're a standard ruler. We can use them basically like grid paper laid over the universe. We just measure how big the grid paper looks far away and how big it looks nearby to see how the universe has expanded from early times until now."

I smiled. It was the quest for H_0 and q_0 again, but with different players. Astronomers had used Type Ia supernovae as standard candles, and they hoped to employ FRBs for the same job. Now Davis and others are trying to see how quickly these bubbles from ancient cosmic sound waves grew at different times in the universe's history. From that, they can better understand dark matter and dark energy, the two things that have contributed the most to what the universe has gotten up to during its 13.7-billion-year life.

"That's the number one aim of this whole survey," Davis said.

To an outsider, it might seem that astronomers default to terms like "dark matter" and "dark energy" because they simply have no clue what's going on. Davis strongly cautioned against that viewpoint.

She motioned to the trees outside her office window. "I can see the leaves of this tree moving around, and they're being blown around by the wind. But if you go back 200 years, we didn't have the periodic table. We didn't know carbon and oxygen and nitrogen, and we didn't have a particle theory of matter. We couldn't explain wind at a fundamental level."

That's where astronomy stands now. Scientists can describe the motions and behaviors of galaxies billions of light-years away quite well.

"Ridiculously well," Davis asserted. "We know in very deep detail the properties of dark energy and dark matter. We just lack the theories to be able to explain them."

By the time the surveys are complete, the team will have a precise picture of the expansion of the universe over the past 10 billion years, nearly the entirety of its lifetime. "We'll have measured it so well that, for me, I think I'll move on to the next thing," Davis said.

"What's that?" I asked.

"What I'm excited about doing is using gravitational waves to measure this," she said.

My eyes must have widened because she laughed. "You didn't see that one coming!"

At the time we spoke, there had been all of 90 gravitational wave events reported, mostly from black hole–black hole mergers, which, I found out, are beautiful standards themselves. They're not typical standard candles like supernovae, with a known luminous brightness, nor are they standard rulers like baryon acoustic oscillations, with a known size. Gravitational waves are instead standard sirens, signals of known "loudness."*

When two compact objects merge, they engage in a very specific dance, the steps unique to the masses of the two objects. Scientists can pick out that dance from the precise way the frequency of the resulting gravitational waves changes. Thanks to the templates that Susan Scott and others tirelessly worked on while LIGO was under construction, astronomers know exactly how "loud" a couple of crashing black holes

---

* Their energy output—luminosity—is well understood, so they are still standard candles in that sense. You know, candles with invisible flames that ripple spacetime.

or neutron stars will be. All Earthlings need to do is measure how loud that signal is when it gets to us, and then we can figure out the exact distance.

"They're the most epic, beautiful standard you can ever imagine," Davis gushed. "The fact that you can know from first principles how bright something is. That's unheard-of in astronomy, right? That's really, really powerful. That's where I'm going to transfer my effort in the next decade."

Because measuring the separations between tens of millions of galaxies wasn't challenging enough, Davis intends to use the next generation of gravitational wave detections to better calibrate the cosmic distances that supernovae are yielding. But this will require fantastic localization of thousands and thousands of gravitational wave sources.

Binary neutron stars would be a great start because they create kilonovae from which astronomers can simultaneously measure redshifts to their host galaxies. But even the so-called dark sirens from merging black holes will be useful in the coming years.

"Ideally, you would like to localize an event down to only one galaxy so you know the host," Davis explained. "But even if you can't, you can localize it down to the 100 galaxies in that region and look at the distribution of those redshifts to figure out the possible host."

As in the case of the baryon acoustic oscillations, no patterns would emerge in the first 100, 1,000, or even 10,000 events. But over time, as gravitational wave astronomers detect sources with the efficiency with which astronomers find other transients, the picture will slowly become clearer.

"It's actually statistically not that hard compared to some of the other stuff we're already doing," she assured me.

I had a tough time envisioning a future with hundreds of thousands of gravitational wave detections when five years of discovery had netted fewer than 100. But then again, Zwicky probably would never have imagined that over 20,000 supernovae would be discovered in the year 2021 alone.

"But you're going to need everyone online," Davis cautioned. "It won't just be a LIGO thing or LIGO-plus-Virgo."

It will take all the gravitational wave observatories around the world with the expertise of thousands of scientists, engineers, and support staff. The good news, according to Davis, is that there will be no need for any instant follow-up work. Scientists can take their time plowing through the statistics, and ultimately this study will help answer some of the biggest questions.

What has the universe been doing for the past 13.7 billion years? And what are its plans for the future?

CHAPTER 25

The Last Hurrah

We are like butterflies who flutter for a day and think
it is forever.

—Carl Sagan, *Cosmos*

There will come, in the deepest of deep futures, a final, almost defiant transient event. Nobody will witness it. Or it might not happen at all if the universe manages to tear itself apart before then or if it turns out that protons have the ability to decay. But it's interesting to speculate what that last event might be.

Here's an incomplete list of what this event most assuredly will *not* be. It will not be the explosive death of a massive star. Those types of stars will all be long dead in an unfathomably ancient, rapidly expanding universe that can barely get two atoms to meet, much less enough atoms to build a massive star.

It will not be any kind of merging event either. All binary systems will have long since coalesced into a single object or been disrupted, the two dance partners gravitationally wrenched apart by a passing third object and cast adrift in a frozen, empty universe.

Supermassive black holes will have swallowed everything they could and collided as often as they could with each other, but there won't be any of those left either.

Instead, what will most likely get the last laugh are the lonely remnants of mediocre stars: white dwarfs. Except by then, they won't even be white. They will have spent the life of the universe cooling and cooling until their temperatures are nothing like the million or so degrees Celsius that they started with. It was this scorching temperature that gave them their intense white glow, a color that persisted even as they cooled to 100,000 degrees Celsius and grew a crystalline interior structure. But as the gigayears passed, they became yellow, then red, and then faded from sight altogether. At the time of their hypothetical swan song, they will be black, no longer radiating at all as they and the universe have reached the coldest possible equilibrium.

This admittedly will all take place a long time from now.

This scenario is so far into the future that each of these objects—however many there are—will occupy in a very real sense its own island universe. These will be nothing like the island universes debated by Curtis and Shapley in the 1920s, constrained by the limited imagination of humans. Instead, these frozen white dwarfs will each enjoy a personal space immeasurably larger than the volume of the currently observable universe. Moreover, each introvert's bubble will be growing exponentially larger, making it literally impossible for any of these objects to have any impact on anything else.

Ever.

And yet, despite the fact that such things have exactly zero chance of ever being observationally verified, this has not stopped people like Matt Caplan of Illinois State University from thinking about them.

"It's just so dang cool!" he exclaimed without a hint of irony that the objects he's hypothesizing are literally at absolute zero. Dang cool, indeed. Sporting dark-rimmed glasses and bushy chaotic hair on which his sunglasses were perched, Caplan had that kid-in-a-candy-shop enthusiasm characteristic of every researcher I've encountered.

He was presenting a lecture remotely from Illinois through a Zoom meeting run by California State University, Long Beach, a colloquium I was viewing while preparing for a hurricane to hit Houston. Such is post-pandemic science communication.

In fact, such is post-pandemic science. A student in an online class at the beginning of the COVID pandemic had noted that the textbook equation for a white dwarf's limiting mass depended on the proportion of electrons. "But what if it's made of something different?"

"It's the sort of thing I'd never even bothered to think about on my own," Caplan admitted.

And so Caplan wondered, what if?

In the current universe, the hypothetical is pointless. White dwarfs aren't made of anything different. They're almost completely carbon and oxygen, or in rare instances oxygen, magnesium, and neon. Their behavior is fairly well understood, and their limiting mass is the same one calculated by Chandrasekhar in 1930: 1.44 times the mass of the Sun.

In the far future, though, dead stellar cores will slowly turn to iron, a transition that changes everything.

"I had the stroke-of-genius movie moment where I ran out of the classroom to write it all down," Caplan said.

"Wasn't this a Zoom class?" I asked.

He conceded, "Okay, that's not what happened, but it can be fun to pretend it was."

So while we're having fun pretending . . .

Imagine the Sun. Now halfway through its hydrogen-fusing life, in about 5 billion years, it will swell to encompass Earth and everything that you ever cared about here. The atoms from your favorite car or cat or garden gnome will still exist, however, and maybe—while we're pretending—those don't all get swept out into space along with the Sun's future planetary nebula. Maybe they settle onto the surface of the dense remnant.

And maybe, because the Sun is in the outskirts of the Milky Way Galaxy, the inevitable collision between our Galaxy and the Andromeda Galaxy will not simply catapult us into their merging supermassive black holes. Maybe in just a few billion years, the Sun's white dwarf, along with the atoms of your favorite car or cat or garden gnome, will be nudged out to a safer distance instead.

And maybe, during all the gravitational jostling of the billions of stars in these merging galaxies, the Sun will win the gravitational lottery. Perhaps through exceedingly improbable, but not impossible channels, it might become paired up with a similar white dwarf. They might gradually spiral together—it's not as though they don't have the time, after all—and combine to form a white dwarf with about 1.16 times the mass of the current Sun.

Over the next many billions of years, the new and improved white dwarf will cool to a frozen black dwarf while the rest of the universe ties up the loose ends. In 100 trillion years, everything that can be processed in the universe will have been processed. Hydrogen fuel will have long been exhausted, and most of the matter in the universe will now be found only in the densest of remnants: black holes, neutron stars, and black dwarfs. There will be no shining stars, no constellations, no glowing gases decorating the night sky.

In 10 million trillion years, galaxies themselves will dissipate. The higher-mass objects will bully the lower-mass objects to leave the galaxies altogether. The black dwarf containing the Sun's remnant, along with the atoms of your favorite cat or car or garden gnome, might be evicted to safety. Over the next million quadrillion years, all remaining galactic orbits will slowly lose energy as they radiate gravitational waves and spiral inexorably toward the super-supermassive black hole at the core.

Now, fast-forward to 10^{50} years in the future, a number so great that we don't have handy words for it. It's 100,000 trillion quadrillion quintillion. The matter in every atom that currently makes up everything you now experience, including this book in your hands and the faintest star in the night sky, is now either part of those scattered supermassive black holes or tied up in the low-mass objects that were kicked out of their dying galaxies.

It's very cold and very dark.

But the universe is still young.

After about 10^{100} years—a googol years—even supermassive black holes with masses of 100 trillion times that of the Sun will have slowly eroded away thanks to an incessant and infinitesimally small trickle of radiation, their lives just a heartbeat in the grand cosmic scheme of eternity. Meanwhile, those previously evicted low-mass objects endure.

Like their former selves, the frozen black dwarfs still have interiors abuzz with electrons that are incessantly doing battle with gravity. The previously agitated carbon and oxygen nuclei have now settled into a crystalline structure, lying dormant in preparation for the last big show.

Now, we wait some more.

At the inconceivably distant future 10^{1100} years—that is a 1 followed by 1,100 zeroes, a number so large that it would take an entire page just to write out the zeroes—from now, things will finally begin to fall into place. The seemingly rigid crystal lattice in the black dwarf will turn out to be not quite so rigid after all. On the scales of the very tiny, particles frequently find themselves teleported to a different place, one that seems at first glance to be out of bounds. Protons do this all the time inside the Sun you see today, showing up beyond the repulsive wall of fellow protons and allowing for the production of sunshine. In the extremely distant future, entire nuclei will ultimately and inevitably do the same inside frozen black dwarfs. Two carbon or oxygen nuclei occupying nearby spots on the crystal latticework will become fused, not because the temperatures are so high, but because the universe simply finds a way if given enough time.

It has at long last been given enough time.

The slow, cold process known as "pycnonuclear fusion" will start where the pressure is the greatest, which is to say in the center of the black dwarf. From there, it will creep outward at a pace that makes glaciers appear impatient. Then, a new pycnonuclear fusion process will create even heavier nuclei, like nickel, each containing 28 protons and 28 neutrons. But this configuration is just a bit off, energetically speaking, so a couple of protons will adjust themselves to make neutrons. To make this happen, the two protons will shed their positive charges as positrons. Now we have iron, which has 26 protons and 30 neutrons. The antimatter counterpart to electrons, the positrons will race out into the buzzing sea of electrons and pick them off, one by one.

The situation seems wholly unfair. The electrons have been dutifully fighting for the past 10^{1100} years to keep gravity from winning. Individually, an electron can't hope to prop up a black dwarf against the crush of gravity, but collectively the trillions upon quadrillions upon quintillions of them have done the job faithfully for more years than can be counted.

Now, though, Caplan said, "it's like knocking out the support beams in the basement of the star." Once securely under the limiting mass of 1.44 solar masses, a black dwarf with a mere 1.35 solar masses, after 10^{1100} long and thankless years, will be unable to prevent collapse. The least massive ones, containing just 16% more mass than the Sun, will manage a good run of about 10^{32000} years before this happens.

It makes no substantive difference if one uses the unit "microseconds" rather than "years." They're both pointlessly short units of time in the grand scheme of things. Even using the entire lifetime of the Sun—10 billion years—as the timekeeping unit merely changes the exponent to 31,990, which is really just 32,000. This is how ridiculously enormous these numbers are, and this is why people have asked Caplan, "Doesn't it make you sad to have an idea about something that you will never, ever, ever observe?"

The universe is filled with such things. The joy, according to Caplan, is in imagining what is possible and seeing where the weird ideas take you.

So maybe, just maybe, in 10^{32000} years or microseconds or solar lifetimes, the remnant of the Sun, having long since merged with another similarly massive stellar core, will be part of a 1.16-solar-mass black dwarf, along with some of the atoms that made up your favorite car or cat or garden gnome.

The electron supports will be eliminated, and gravity will begin to win for the last time. Once the collapse begins, the elderly black dwarf is done for. It's the story of a massive star core-collapse supernova all over again, but this time it will happen in the empty stellar zombie realm of the deepest, darkest future. The remaining electrons, succumbing to the new intense pressure, will combine with protons and spit out the universe's last flood of neutrinos, and the floor will drop out from underneath the rest of the star. A full foe of energy will blast the rest of the star into space in the universe's last transient event ever.

No observatory will ever detect light or neutrinos, cosmic rays or gravitational waves from this event.

Nothing new will ever form from its remnants.

But maybe, just maybe, at least part of our mediocre star will have the last word.

Ephemera

We are tiny, short-lived beings looking in wonder at the energetic and mysterious things that go bump in the universe. That we can make any sense of them at all is astonishing. But that we try is a given. After all, they are our cousins. Always changing. Under pressure. Sometimes caving in. Sometimes exploding. Frequently colliding. Following patterns laid down long ago. Frantic, tempestuous, luminous, self-destructive, creative.

Brilliant.

Even if for just an instant.

ACKNOWLEDGMENTS

While I was writing this book, a tweet by Tim Urban (@waitbutwhy) on 20 May 2021 began circulating through my social media feeds. "Stars are basically the immediate after-effects of the Big Bang," it read. "A one-second sizzle of brightness before settling into an endless era of darkness. We live in that one bright second."

There have been so many bright seconds, fortuitous transient events, and happenstance collisions that were made brighter by the immeasurable generosity of astronomical researchers around the world. Every researcher I chatted with and every observatory I visited told a rich, wondrous story deserving of so much more time in the spotlight. Whittling those countless hours to a page, a paragraph, or a passing mention—or even less—feels unfair. To everyone who gave so freely of their time, I owe a massive debt of gratitude.

I would especially like to honor the memories of Alison Doane (Harvard University), who revealed the glass universe to me, and Kevin Hurley (University of California, Berkeley), who shared with me the thrill of chasing gamma ray bursts. I hope I have done their work justice.

Special thanks go to Sarah Burke-Spolaor (West Virginia University), Jeff Cooke (Swinburne University of Technology), Tamara Davis (University of Queensland), Orsola De Marco (Macquarie University), Emmanuel Fonseca (West Virginia University), Duane Hamacher (University of Melbourne), George Hobbs (CSIRO Space and Astronomy), Rob Hollow (CSIRO Space and Astronomy), Simon Johnston (CSIRO Space and Astronomy), Amanda Karakas (Monash University),

Paul Lasky (Monash University), Duncan Lorimer (West Virginia University), Ilya Mandel (Monash University), Maura McLaughlin (West Virginia University), Smadar Naoz (UCLA), Benjamin Pope (University of Queensland), Nikhil Sarin (Stockholm University), John Sarkissian (CSIRO Space and Astronomy), Susan Scott (Australian National University), Nathan Smith (University of Arizona), Nicholas Suntzeff (Texas A&M University), Lawrence Toomey (CSIRO Space and Astronomy), Sara Webb (Swinburne University of Technology), and finally Matt Caplan (Illinois State University), who has the added distinction of providing the largest number in this book.

Plenty of other people generously gave their time and expertise to help me understand the wide range of topics. I would like to offer heartfelt thanks to Corey Austin (LIGO Livingston), John Barentine (International Dark-Sky Association), Kathy Holt (LIGO Livingston), Lisa Koerner (University of Houston), Ryan Lynch (Green Bank Observatory), Jill Malusky (Green Bank Observatory), Richard McDermid (Macquarie University), Oliver Roberts (Universities Space Research Association), and Andy Seymour (Green Bank Observatory).

An earlier version of portions of chapters 6 and 24 was published in *Astronomy* magazine and is used here with permission from editor Dave Eicher. Part of chapter 21 was previously published in *Mercury* magazine and is used with the kind permission of editor Liz Kruesi.

I owe a debt of gratitude to everyone who read and provided feedback on any portion of this book while it was in its larval phase, including everyone recognized above. Special thanks go to Holly Jean Richard, who helped me find my voice early on. Closer to home, my daughter, Megan, got to experience the entire sausage-making process and even learned a few cool things along the way. Sam Houston State University student Charise Lincoln not only provided feedback, but also was instrumental in organizing the finished product. I also want to thank my father, Don R. James, who spotted nearly every typo in the first draft and who really wanted me to find a way to work the name Willem Dafoe into the text. There. I've done it.

Of course, none of this would have been possible without both financial support and uninterrupted blocks of writing time. The former was provided by the Alfred P. Sloan Foundation Program for the Public Understanding of Science, Technology, and Economics, which funded much of the travel for the research and writing of the book. The latter came from Sam Houston State University in the form of a sabbatical that was thoroughly encouraged by my administration and colleagues. I am beyond fortunate to work with people who are so continually supportive of my academic pursuits.

Finally, I would like to thank my editor, Tiffany Gasbarrini, and everyone at Johns Hopkins University Press for their confidence in me and in this project.

One bright second, indeed!

BIBLIOGRAPHY

CHAPTER 1. Catching Cosmic Fireflies

Aristotle. *On the Heavens*. Translated by W. K. C. Guthrie. Cambridge, MA: Harvard University Press, 1939.

Fredrick, Serena. "The Sky of Knowledge: A Study of the Ethnoastronomy of the Aboriginal People of Australia." Master's thesis, University of Leicester, Leicester, England, 2008.

Goldstein, Bernard R. "Evidence for a Supernova of AD 1006." *Astronomical Journal* 70, no. 1 (1965): 105–114.

Hamacher, Duane. "Are Supernovae Recorded in Indigenous Astronomical Traditions?" *Journal of Astronomical History and Heritage* 17, no. 2 (2014): 161–170.

Katsuda, Satoru. "Supernova of 1006 (G327.6 + 14.6)." In *Handbook of Supernovae*, edited by Athem W. Alsabti and Paul Murdin. Cham, Switzerland: Springer, 2017. https://doi.org/10.1007/978-3-319-20794-0_45-2.

Norris, Cilla, and Ray Norris. *Emu Dreaming: An Introduction to Australian Aboriginal Astronomy*. Sydney, Australia: Emu Dreaming, 2014.

Sagan, Carl. *Cosmos*. New York: Random House, 1980.

Shu Wang, et al. "The Mid-Infrared Extinction Law and Its Variation in the Coalsack Nebula." *Astrophysical Journal* 773, no. 1 (2013): 30.

Stephenson, F. R. "Revised Catalogue of Pre-Telescopic Galactic Novae and Supernovae." *Quarterly Journal of the Royal Astronomical Society* 17 (1976): 121.

———. "SN 1006: The Brightest Supernova." *Astronomy and Geophysics* 51, no. 5 (2010): 5.27–5.32.

Stephenson, F. R., and D. A. Green. *Historical Supernovae and Their Remnants*. Oxford: Clarendon, 2002.

CHAPTER 2. Out of the Question

Clerke, Agnes M. *The System of the Stars*. London, England: Longmans, Green, 1890.

Curtis, Heber B. "The Distances of the Stars." *Publications of the Astronomical Society of the Pacific* 23, no. 136 (1911): 349–355.

Harvard and Smithsonian Center for Astrophysics. "Virtual Tours of the Great Refractor, Plate Stacks Now Available." Last modified 2 April 2022. https://www.cfa .harvard.edu/news/virtual-tours-great-refractor-plate-stacks-now-available?fbclid =IwAR2VTrA9wR-3bmugjapRU2zpmuz1gDkAFVrmgN6EC7aJO8wzL9IcSLGnAz0.

Hubble, Edwin P. "Cepheids in Spiral Nebulae." *Popular Astronomy* 32 (1925): 252–255.

Jones, Bessie Zaban, and Lyle Gifford Boyd. *The Harvard College Observatory: The First Four Directorships, 1839–1919.* Cambridge, MA: Belknap, 1971.

Jones, K. G. "S Andromedae, 1885: An Analysis of Contemporary Reports and a Reconstruction." *Journal for the History of Astronomy* 6 (1885): 27–40.

Leavitt, Henrietta S. "1777 Variables in the Magellanic Clouds." *Annals of the Astronomical Observatory of Harvard College* 60, no. 3 (1908): 87–108.

Leavitt, Henrietta S., and Edward C. Pickering. "Periods of 25 Variable Stars in the Small Magellanic Cloud." *Harvard College Observatory Circular* 172 (1912): 1–3.

Payne-Gaposchkin, Cecilia. *An Autobiography and Other Recollections.* 2nd ed. Cambridge: Cambridge University Press, 1996.

Shapley, Harlow. "On the Existence of External Galaxies." *Journal of the Royal Astronomical Society of Canada* 12 (1919): 438–446.

Sobel, Dava. *The Glass Universe.* New York: Viking, 2016.

Wolbach Library. "Project Phaedra: Preserving Harvard's Early Data and Research in Astronomy." https://library.cfa.harvard.edu/project-phaedra.

CHAPTER 3. Putting the "Super" in Supernova

Baade, W., and F. Zwicky. "Cosmic Rays from Super-Novae." *Proceedings of the National Academy of Sciences* (1934): 259–263. https://doi.org/10.1073/pnas.20.5.259.

———. "On Super-Novae." *Proceedings of the National Academy of Sciences* (1934): 254–259. https://doi.org/10.1073/pnas.20.5.254.

Johnson, John, Jr. *Zwicky: The Outcast Genius Who Unmasked the Universe.* Cambridge, MA: Harvard University Press, 2019. https://doi.org/10.4159 /9780674242616.

Mancuso, Richard V., and Kevin R. Long. "The Astro-Blaster." *Physics Teacher* 32 (1995): 358. https://doi.org/10.1119/1.2344238.

Zwicky, F. "Characteristic Temperatures in Super-Novae." *Proceedings of the National Academy of Sciences of the United States of America* 22, no. 8 (1936): 557–561. https://doi.org/10.1073/pnas.20.5.259.

———. "A Super-Nova in NGC 4157." *Publications of the Astronomical Society of the Pacific* 49, no. 289 (1937): 204–206.

CHAPTER 4. Star-Shattering Energy

Atmanspacher, Harald, and Hans Primas. "Pauli's Ideas on Mind and Matter in the Context of Contemporary Science." *Journal of Consciousness Studies* 13 (2006): 5–50.

Branch, David, and J. Craig Wheeler. *Supernova Explosions*. Berlin, Germany: Springer-Verlag, 2017.

Einstein, Albert. "Does the Inertia of a Body Depend upon Its Energy-Content?" 1905. https://www.fourmilab.ch/etexts/einstein/E_mc2/e_mc2.pdf.

Hubble, Edwin P. "Novae or Temporary Stars." *Astronomical Society of the Pacific Leaflets* 1, no. 14 (1928): 55–58.

Mayall, N. U. "The Crab Nebula, a Probable Supernova." *Astronomical Society of the Pacific Leaflets* 3, no. 119 (1939): 145–154.

CHAPTER 5. The Search for Smoking Guns

Baade, W., and R. Minkowski. "On the Identification of Radio Sources." *Astrophysical Journal* 119 (1954): 215–231.

Gardner, F. F., and D. K. Milne. "The Supernova of 1006 A.D." *Astronomical Society* 70, no. 9 (1965): 754.

Hazard, Cyril, et al. "The Sequence of Events That Led to the 1963 Publications in Nature of 3C 273, the First Quasar and the First Extragalactic Radio Jet." *Publications of the Astronomical Society of Australia* 35 (2018). https://doi.org/10.1017/pasa .2017.62.

Jansky, C. M. "The Beginnings of Radio Astronomy." *American Scientist* 45, no. 1 (1957): 5–12.

Jansky, Karl G. "Electrical Phenomena That Apparently Are of Interstellar Origin." *Popular Astronomy* 41 (1933): 548–555.

Schmidt, Maarten. "The Discovery of Quasars." *Proceedings of the American Philosophical Society* 155, no. 2 (2011): 142–146.

CHAPTER 6. Detecting Cosmic Heartbeats

Baym, Gordon, and Christopher Pethick. "Neutron Stars." *Annual Review of Nuclear Science* 25, no. 1 (1975): 27–77. https://doi.org/10.1146/annurev.ns.25.120175.000331.

Bell Burnell, S. J. "Petit Four: After Dinner Speech, Mullard Space Science Laboratory." *Annals of the New York Academy of Sciences* 302 (1977): 685–689.

Galloway, Duncan. "Explainer: What Is a Neutron Star?" *Conversation*, 31 August 2015. https://theconversation.com/explainer-what-is-a-neutron-star-29341.

Hewish, A., et al. "Observation of a Rapidly Pulsating Radio Source." *Nature* 217 (1968): 709–713. https://doi.org/10.1038/217709a0.

Klebesadel, Ray W., Ian B. Strong, and Roy A. Olson. "Observations of Gamma-Ray Bursts of Cosmic Origin." *Astrophysical Journal* 182 (1973): L85–L88.

Manchester, R. N. "Finding Pulsars at Parkes." *Publications of the Astronomical Society of Australia* 18 (September 2000): 1–16. https://doi.org/10.48550/arXiv.astro-ph /0009405.

Michaelis, Anthony. "Space Signals May Be from Intelligent Being: Pulsating Star Traced." *Daily Telegraph*, March 5, 1968.

Morton, Donald C. "Neutron Stars as X-Ray Sources." *Nature* 201, no. 4926 (March 1964): 1308–1309.

Nervosa, Alex. "Detecting Pulsars." Astronomy Online. http://astronomyonline.org /Stars/Pulsars.asp.

Ott, Christian D., et al. "The Spin Periods and Rotational Profiles of Neutron Stars at Birth." *Astrophysical Journal Supplement Series* 164, no. 1 (May 2006): 130–155. https://doi.org/10.1086/500832.

Pais, Helena, and Jirina R. Stone. "Exploring the Nuclear Pasta Phase in Core-Collapse Supernova Matter." *Physical Review Letters* 109 (October 2012): 151101. https:// journals.aps.org/prl/abstract/10.1103/PhysRevLett.109.151101.

Proust, Marcel. *The Captive*. Vol. 5 of *Remembrance of Things Past*. Trans. C. K. Scott Moncrieff. New York: Random House, 1934.

Reisenegger, Andreas. "Origin and Evolution of Neutron Star Magnetic Fields." Paper presented at International Workshop on Strong Magnetic Fields and Neutron Stars, Havana, Cuba, 7–12 April 2003. https://doi.org/10.48550/arXiv.astro-ph/0307133.

CHAPTER 7. Stellar Arrhythmia

Andersson, N. "A Superfluid Perspective on Neutron Star Dynamics." *Universe* 7, no. 1 (January 2021): 1–18. https://doi.org/10.48550/arXiv.2103.10218.

Backer, D. C., et al. "A Millisecond Pulsar." *Nature* 300, no. 5893 (December 1982): 615–618.

Baym, Gordon, and David Pines. "Neutron Starquakes and Pulsar Speedup." *Annals of Physics* 66 (1971): 816–835.

Condon, James J., and Scott M. Ransom. *Essential Radio Astronomy*. Princeton, NJ: Princeton University Press, 2017.

Haskell, Brynmor, and Armen Sedrakian. "Superfluidity and Superconductivity in Neutron Stars." *Astrophysics and Space Science Library* 457 (2017): 401–454. https://doi.org/10.48550/arXiv.1709.10340.

Ho, Wynn C., et al. "Pinning Down the Superfluid and Measuring Masses Using Pulsar Glitches." *Science Advances* 1, no. 9 (October 2015). https://doi.org/10.1126/sciadv .1500578.

Johnston, Hamish. "Pulsar Glitch Suggests Superfluid Layers Lie within Neutron Star." *Physics World*, 15 August 2019.

Johnston, Simon, and Aris Karastergiou. "Pulsar Braking and the P-Dot Diagram." *Monthly Notices of the Royal Astronomical Society* 467, no. 3 (February 2017): 3493–3499. https://doi.org/10.48550/arXiv.1702.03616.

Kohler, Susanna. "More Insight into Neutron Star Interiors." American Astronomical Society, *Nova*, 15 September 2021. https://aasnova.org/2021/09/15/more-insight -into-neutron-star-interiors.

"Legacy Discoveries." National Science Foundation: The Arecibo Observatory. https://www.naic.edu/ao/legacy-discoveries.

Molnar, L., et al. "KIC 9832227: A Red Nova Precursor." *Bulletin of the American Astronomical Society* 229 (2017): abstract no. 417.04.

Radhakrishnan, V. "Fifteen Months of Pulsar Astronomy." *Publications of the Astronomical Society of Australia* 1, no. 6 (1969): 254–263. https://doi.org/10.1017/S1323358000011826.

Radhakrishnan, V., and G. Srinivasan. "On the Origin of the Recently Discovered Ultra-Rapid Pulsar." *Current Science* 51, no. 23 (December 1982): 1096–1099.

Rencoret, Javier A., Claudia Aguilera-Gómez, and Andreas Reisenegger. "Revisiting Neutron Starquakes Caused by Spin-Down." *Astronomy and Astrophysics* 654, no. A47 (August 2021): 1–17. https://doi.org/10.1051/0004-6361/202141499.

CHAPTER 8. (Almost) No Star Is an Island

Cadelano, M., et al. "Discovery of Three New Millisecond Pulsars in Terzan 5." *Astrophysical Journal Letters* 855, no. 2 (March 2018): 1–7. https://doi.org/10.3847/1538-4357/aaac2a.

"Castor, the 6-Star System: A3 Poster." Jet Propulsion Laboratory: California Institute of Technology, 10 August 2012. https://www.jpl.nasa.gov/infographics/castor-the-6-star-system-a3-poster.

"Exploding Binary Stars Will Light Up the Sky in 2022." *Futurism*, 28 November 2017. https://futurism.com/exploding-binary-stars-light-sky-2022.

Guszejnov, David, and Mike Grudic. "Star Formation in Gaseous Environments." STARFORGE Project, 26 October 2021. https://www.starforge.space/index.html.

"Hubble Discovers Rare Fossil Relic of Early Milky Way." European Space Agency, 7 September 2016. https://esahubble.org/news/heic1617.

Kennedy, M. R., et al. "Kepler K2 Observations of the Transitional Millisecond Pulsar PSR J1023 + 0038." *Monthly Notices of the Royal Astronomical Society* 477, no. 1 (2018): 1120–1132. https://doi.org/10.48550/arXiv.1310.0544.

Kinemuchi, Karen. "To Pulsate or to Eclipse? Status of KIC 9832227 Variable Star." Paper presented at 40 Years of Variable Stars: A Celebration of Contributions by Horace A. Smith, October 2013, Michigan State University, Lansing. https://doi.org/10.48550/arXiv.1310.0544.

Legrand, Michel, with Alan Bergman and Marilyn Bergman. "The Windmills of Your Mind." Reprise Records, 1968. RS.20758.

"NASA: Fermi Catches a 'Transformer' Pulsar." YouTube, 22 July 2014. https://www.youtube.com/watch?v=Hn5RJ2PN718.

Parks, Jake. "Two Stars Won't Collide into a Red Nova in 2022 after All." *Discover Magazine*, 7 September 2018. https://www.discovermagazine.com/the-sciences/two-stars-wont-collide-into-a-red-nova-in-2022-after-all.

Socia, Quetin J., et al. "KIC 9832227: Using Vulcan Data to Negate the 2022 Red Nova Merger Prediction." *Astrophysical Journal Letters* 864, no. L32 (September 2018): 1–12. https://doi.org/10.48550/arXiv.1809.02771.

Twain, M., and G. Scharnhorst. *Mark Twain: The Complete Interviews*. Tuscaloosa: University of Alabama Press, 2006.

Wheeler, J. Craig. *Cosmic Catastrophes: Exploding Stars, Black Holes, and Mapping the Universe*, 2nd ed. Cambridge: Cambridge University Press, 2007.

CHAPTER 9. The Making of a Superstar

Sterken, C. "The Drama of η Carinae." *Information Bulletin on Variable Stars* 5000, no. 1 (December 2000).

Thackeray, A. D. "The Southern Skies: An Underworked Goldmine." *Monthly Notes of the Astronomical Society of Southern Africa* 21, no. 96 (1962): 96–99.

CHAPTER 10. Cloudy with a Chance of Neutrinos

"About: Kamioka Observatory." Kamioka Observatory. https://www-sk.icrr.u-tokyo.ac.jp /en/about/?doing_wp_cron=1653003252.8055830001831054687500.

Al-Kharusi, S., et al. "SNEWS 2.0: A Next-Generation Supernova Early Warning System for Multi-Messenger Astronomy." *New Journal of Physics* 23 (March 2021): 1–34. https://doi.org/10.1088/1367-2630/abde33.

Bionta, R. M., et al. "Observation of a Neutrino Burst in Coincidence with Supernova 1987A in the Large Magellanic Cloud." *Physical Review Letters* 58, no. 14 (April 1987): https://doi.org/10.1103/PhysRevLett.58.1494.

Boyd, Steve. "Neutrino Detectors." University of Warwick, Coventry, England, 2016. https://warwick.ac.uk/fac/sci/physics/staff/academic/boyd/stuff/px435/lec _neutrinodetectors_2016.pdf.

Burnham, Robert, Jr. *Burnham's Celestial Handbook*. Vol. 2. New York: Dover, 1966.

Central Bureau for Astronomical Telegrams. "Supernova 1987A in the Large Magellanic Cloud." *International Astronomical Union Circular* 4316 (24 February 1987).

Cigan, P., et al. "High Angular Resolution ALMA Images of Dust and Molecules in the SN 1987A Ejecta." *Astrophysical Journal* 886, no. 1 (November 2019): 51. https://doi .org/10.3847/1538-4357/ab4b46.

Greco, E., et al. "Indication of a Pulsar Wind in the Hard X-Ray Emission from SN 1987A." *Astrophysical Journal* 908 (February 2021): L45–L51.

Koshiba, M. "Observation of Neutrinos from SN 1987A by Kamiokande-II." *Proceedings of ESO Workshop on the SN 1987A* (July 1987): 221.

Menon, Athira, Victor Utrobin, and Alexander Heger. "Explosions of Blue Supergiants from Binary Mergers for SN 1987A." *Monthly Notices of the Royal Astronomical Society* 482, no. 1 (January 2019): 438–452. https://doi.org/10.1093/mnras /sty2647.

Nakamura, Ko. "Core-Collapse Simulation of SN 1987A Binary Progenitor and Its Multimessenger Signals." *Journal of Physics: Conference Series* 2156 (2021): 1–4. https://doi.org/10.1088/1742-6596/2156/1/012232.

SNEWS. "The Supernova Early Warning System." https://snews2.org.

"Supernova Neutrinos." All Things Neutrino. https://neutrinos.fnal.gov/sources
/supernova-neutrinos.

Suzuki, A. "The 20th Anniversary of SN1987A." *Journal of Physics Conference Series* 120
(2008): 5. https://doi.org/doi:10.1088/1742-6596/120/7/072001.

CHAPTER 11. Not "The End"

Baglin, A. "White Dwarfs and Type I Supernovae." *Astrophysical Letters* 1 (1968): 143.

Cappellaro, E. "Seventy Years of Supernova Searches." In *1604–2004: Supernovae as
Cosmological Lighthouses*, edited by M. Turatto et al. San Francisco: Astronomical
Society of the Pacific, 2005. https://articles.adsabs.harvard.edu/full/2005ASPC..342
...71C.

Chandrasekhar, S. "The Maximum Mass of Ideal White Dwarfs." *Astrophysical Journal*
74 (July 1931): 81–82.

De Kishlay, et al. "A Population of Heavily Reddened, Optically Missed Novae from
Palomar Gattini-IR: Constraints on the Galactic Nova Rate." *Astrophysical Journal* 912,
no. 1 (April 2021): 1–20. https://doi.org/10.48550/arXiv.2101.04045.

Eddington, Arthur S. *Stars and Atoms*. Oxford: Clarendon, 1927.

Einstein, A. "The Foundation of the General Theory of Relativity." *Annalen der Physik*
49 (1916): 769–822.

"First Discovery of a Binary Companion for a Type Ia Supernova." Harvard and
Smithsonian Center for Astrophysics, 22 March 2016. https://www.cfa.harvard.edu
/news/first-discovery-binary-companion-type-ia-supernova.

Hamuy, Mario, et al. "The Hubble Diagram of the Calán/Tololo Type Ia Supernovae and
the Value of H_0." *Astronomical Journal* 112 (December 1996): 1–32. https://doi.org
/10.1086/118191.

Kare, Jordin, et al. "The Berkeley Automated Supernova Search." Paper presented at
North American Treaty Organization Advanced Study Institute, Cambridge, 28
June–10 July 1981. https://doi.org/10.1007/978-94-009-7876-8_20.

Kennefick, D. "Einstein versus the *Physical Review*." *Physics Today* 58, no. 9 (2005): 43.
https://doi.org/10.1063/1.2117822.

Kowal, C. T. "Absolute Magnitudes of Supernovae." *Astronomical Journal* 73, no. 10
(December 1968): 1021–1024.

Kowal, C. T., and W. L. W. Sargent. "Supernovae Discovered since 1885." *Astronomical
Journal* 76, no. 9 (November 1971): 756–764.

Liu, D., B. Wang, and Z. Han. "The Double-Degenerate Model for the Progenitors of
Type Ia Supernovae." *Monthly Notices of the Royal Astronomical Society* 473, no. 4
(October 2017): 5352–5361. https://doi.org/10.1093/mnras/stx2756.

Milne, E. A. "Stellar Structure and the Origin of Stellar Energy." *Nature* 127, no. 3192
(January 1931): 16–27. https://doi.org/10.1038/127016a0.

Minkowski, R. "Spectra of Supernovae." *Publications of the Astronomical Society of the
Pacific* 53, no. 314 (August 1941): 224–225.

Misner, C. W., et al. *Gravitation*. San Francisco, CA: Freeman, 1973.

Perlmutter, S., et al. "Measurements of O and L from 42 High-Redshift Supernovae." *Astrophysical Journal* 517, no. 2 (June 1999): 565–586. https://doi.org/10.1086/307221.

Sandage, A. "Cosmology: A Search for Two Numbers." *Physics Today* 23, no. 2 (1970): 34. https://doi.org/10.1063/1.3021960.

Schmidt, Brian P., et al. "The High-Z Supernova Search: Measuring Cosmic Deceleration and Global Curvature of the Universe Using Type Ia Supernovae." *Astrophysical Journal* 507, no. 1 (November 1998): 46–63. https://doi.org/10.1086/306308.

Tremble, Virginia. "Supernovae. Part I: The Events." *Review of Modern Physics* 54, no. 4 (October 1982): 1185. https://doi.org/10.1103/RevModPhys.54.1183.

CHAPTER 12. Collision Course

Hulse, R. A., and J. H. Taylor. "Discovery of a Pulsar in a Close Binary System." *Bulletin of the American Astronomical Society* 6 (September 1974): 453.

Kennefick, Daniel. "Einstein versus the Physical Review." *Physics Today* 58, no. 9 (September 2005): 43. https://doi.org/10.1063/1.2117822.

"The Nobel Prize in Physics 1993: Press Release." Nobel Prize, 13 October 1993. https://www.nobelprize.org/prizes/physics/1993/press-release.

Plait, Phil. "Binary Pulsar Puts Einstein to the Test . . . and He Passes. Relatively Speaking." *Syfy Wire*, 15 December 2021. https://www.syfy.com/syfy-wire/bad-astronomy-orbiting-pulsars-test-relativity.

Shao, Lijing. "General Relativity Withstands Double Pulsar's Scrutiny." *Physics* 14, no. 173 (December 2021). https://physics.aps.org/articles/v14/173.

Taylor, J. H., et al. "Further Observations of the Binary Pulsar PSR 1913 + 16." *Astrophysical Journal* 206 (May 1976): L53–L58.

Taylor, J. H., and J. M. Weisberg. "A New Test of General Relativity: Gravitational Radiation and the Binary Pulsar PSR 1913 + 16." *Astrophysical Journal* 253 (February 1982): 908–920. https://doi.org/10.1086/159690.

Thorne, K. S. "Relativistic Gravitational Effects in Pulsars." *Comments on Astrophysics and Space Physics* 1 (1969): 12–18.

CHAPTER 13. Fallen Stars

Adams, S. M., et al. "The Search for Failed Supernovae with the Large Binocular Telescope: Confirmation of a Disappearing Star." *Monthly Notices of the Royal Astronomical Society* 488, no. 4 (2017): 4968–4981.

Korobkin, Oleg, et al. "Gamma Rays from Kilonova: A Potential Probe of r-Process Nucleosynthesis." *Astrophysical Journal* 889, no. 2 (February 2020): 168. https://doi.org/10.3847/1538-4357/ab64d8.

Sarin, Nikhil. "The Observational Signatures of Nascent Neutron Stars." PhD thesis, Monash University, 2021. https://doi.org/10.26180/16766923.v1.

CHAPTER 14. Don't Blink

Aptekar, R. L., et al. "General Properties of Recurrent Bursts from SGRs." *Memorie della Societa Astronomica Italiana* 73 (2002): 485–490.

Berman, Bob. "Weird Object: Magnetar SGR 1806-20." *Astronomy*, 4 September 2015. https://astronomy.com/magazine/weirdest-objects/2015/09/17-magnetar-sgr-1806-20.

Blaes, O., et al. "Slowly Accreting Neutron Stars and the Origin of Gamma-Ray Bursts." *Astrophysical Journal* 363 (November 1990): 612–627.

Central Bureau for Astronomical Telegrams. "IAUC 3356: Gamma-Ray Burst 79-03-05." 11 May 1979. http://www.cbat.eps.harvard.edu/iauc/03300/03356 .html#Item1.

Curtis, Heber D. "Descriptions of 762 Nebulae and Clusters Photographed with the Crossley Reflector." *Publications of the Lick Observatory* 13 (1918): 9–42.

Duncan, Robert. "'Magnetars,' Soft Gamma Repeaters & Very Strong Magnetic Fields." March 2003, University of Texas, Austin. https://web.archive.org/web/20130517180957 /http:/solomon.as.utexas.edu/~duncan/magnetar.html.

Fishman, G. J., et al. "Overview of Observations from BATSE on the Compton Observatory." *Astronomy and Astrophysics, Supplemental Series* 97 (January 2003): 17–20.

Hurley, K., et al. "Upper Limits to the Soft Gamma Repeater Population from the LILAS Experiment aboard the PHOBOS Mission." *Astrophysical Journal* 423 (March 1994): 709. https://doi.org/10.1086/173849.

Kouveliotou, C. "BATSE Results on Observational Properties of Gamma-Ray Bursts." *Astronomy and Astrophysics, Supplemental Series* 97, no. 2 (June 1994): 637–642.

Laros, J. G., et al. "A New Type of Repetitive Behavior in a High-Energy Transient." *Astrophysical Journal Letters* 320 (September 1987): L111–L115.

Piran, Tsvi. "Gamma-Ray Bursts and Binary Neutron Star Mergers." In *Compact Stars in Binaries: Proceedings from IAU Symposium, The Hague, Netherlands* (1994), 1–13. https://doi.org/10.48550/arXiv.astro-ph/9412098.

———. "γ-Ray Bursts and Neutron Star Mergers." *Proceedings of the Lanczos Centenary* (1994): 1–25. https://doi.org/10.48550/arXiv.astro-ph/9405006.

Van den Bergh, S. "The Origin of Cosmic Gamma-Ray Bursts." *Astrophysics and Space Science* 97, no. 2 (December 1983): 385–368. https://doi.org/10.1007/BF00653494.

Weekes, Trevor C. "Klebesadel, Strong, & Olson's Discovery of Gamma-Ray Bursts." *Astrophysical Journal, Centennial Issue* 525C (November 1999): 1218.

CHAPTER 15. Point Blank

Allen, Jonathan. "Clue to an Ancient Cosmic-Ray Event?" *Nature* 486 (June 2012): 473. https://doi.org/10.1038/486473e.

Carrington, R. C. "Description of a Singular Appearance Seen in the Sun on September 1, 1859." *Monthly Notices of the Royal Astronomical Society* 20 (November 1859): 13–15. https://doi.org/10.1093/mnras/20.1.13.

Choi, Charles Q. "Real Death Star Could Strike Earth." *Space*, 9 March 2008. https://www.space.com/5081-real-death-star-strike-earth.html.

"Cosmic Explosion among the Brightest in Recorded History." NASA, 18 February 2005. https://www.nasa.gov/vision/universe/watchtheskies/swift_nsu_0205.html.

Dar, Arnon, and A. De Rujula. "The Threat to Life from Eta Carinae and Gamma Ray Bursts." *Astrophysics and Gamma Ray Physics in Space* 24 (2002): 513–523. https://doi.org/10.48550/arXiv.astro-ph/0110162.

Hadhazy, A. "A Scary 13th: 20 Years Ago Earth Was Blasted with a Massive Plume of Solar Plasma." *Scientific American*, 13 March 2009. https://www.scientificamerican.com/article/geomagnetic-storm-march-13-1989-extreme-space-weather/.

Hapgood, M. "The Great Storm of May 1921: An Exemplar of a Dangerous Space Weather Event." *Space Weather* 17 (2019): 950–975. https://doi.org/10.1029/2019SW002195.

Kaspi, Victoria. "Fast Radio Bursts." 20th Hintze Lecture, Oxford University, December 2020. https://www.youtube.com/watch?v=7djjvnpY68k&t=2936s.

Levan, Andrew, Paul Crowther, and Sung-Chul Yoon. "Gamma-Ray Burst Progenitors." *Space Science Reviews* 202 (November 2016): 33–78. https://doi.org/10.1007/s11214-016-0312-x.

Love, Jeffrey J. "Extreme-Event Magnetic Storm Probabilities Derived from Rank Statistics of Historical Dst Intensities for Solar Cycles 14–24." *Space Weather* 19, no. 4 (April 2021). https://doi.org/10.1029/2020SW002579.

Lovett, R. "Mysterious Radiation Burst Solved?" *Nature*, 27 June 2012. https://doi.org/10.1038/nature.2012.10898.

"LSU Astronomer and NASA Missions Unmask Cosmic Eruptions in Nearby Galaxies." LSU Department of Physics and Astronomy. https://www.lsu.edu/physics/news/2021/eric_burns_nasa_magnetar.php.

Mazets, E. P., et al. "Catalog of Cosmic Gamma-Ray Bursts from the KONUS Experiment Data: Preface." *Astrophysics and Space Science* 80 (November 1981).

Melott, A., et al. "Did a Gamma-Ray Burst Initiate the Late Ordovician Mass Extinction?" *International Journal of Astrobiology* 3, no. 1 (January 2004): 55–61. https://doi.org/10.1017/S1473550404001910.

Miyake, Fusa, et al. "A Signature of Cosmic-Ray Increase in AD 774–775 from Tree Rings in Japan." *Nature* 486 (June 2012): 240–242.

"Multiwavelength Astronomy: The First Optical Counterpart." eCUIP: The Digital Library Project. https://ecuip.lib.uchicago.edu/multiwavelength-astronomy/gamma-ray/impact/07.html.

Namekata, Kosuke, et al. "Optical and X-Ray Observations of Stellar Flares on an Active M Dwarf AD Leonis with the Seimei Telescope, SCAT, NICER, and OISTER." *Publications of the Astronomical Society of Japan* 73, no. 2 (April 2022): 485. https://doi.org/10.1093/pasj/psab013.

Parley, Daniel. "Black Holes: We Think We've Spotted the Mysterious Birth of One." *Conversation*, 12 January 2002. https://theconversation.com/black-holes-we-think -weve-spotted-the-mysterious-birth-of-one-174726.

Phillips, Tony. "The Day Earth Lost Half Its Satellites (Halloween Storms 2003)." 28 October 2021. https://spaceweatherarchive.com/2021/10/28/the-day-earth-lost-half -its-satellites-halloween-storms-2003/.

Plait, Phil. "Anniversary of a Cosmic Blast." *Slate*, 27 December 2012. https://slate.com /technology/2012/12/cosmic-blast-magnetar-explosion-rocked-earth-on-december -27-2004.html.

"Soho's Pioneering 25 Years in Orbit." European Space Agency, 22 February 2020. https://www.esa.int/Science_Exploration/Space_Science/SOHO_s_pioneering_25 _years_in_orbit.

Tuthill, Peter. "WR 104: Technical Questions." University of Sydney, Australia. http://www.physics.usyd.edu.au/~gekko/pinwheel/tech_faq.html.

Wang, F. Y., et al. "Consequences of Energetic Magnetar-Like Outbursts of Nearby Neutron Stars: 14C Events and the Cosmic Electron Spectrum." *Astrophysical Journal* 887, no. 2 (December 2019). https://doi.org/10.3847/1538-4357/ab55db.

CHAPTER 16. Cats, Rats, and Fantastic Beasts

Burke-Spolaor, Sarah. "Multiple Messengers of Fast Radio Bursts." *Nature Astronomy* 2 (November 2018): 845–848. https://doi.org/10.48550/arXiv.1811.00194.

Burke-Spolaor, Sarah, et al. "Radio Bursts with Extragalactic Spectral Characteristics Show Terrestrial Origins." *Astrophysical Journal* 727, no. 1 (January 2011): 1–5. https://doi.org/10.1088/0004-637X/727/1/18.

Keane, E. F., et al. "The Survey for Pulsars and Extragalactic Radio Bursts: I. Survey Description and Overview." *Monthly Notices of the Royal Astronomical Society* 473, no. 1 (January 2018): 116–135. https://doi.org/10.1093/mnras/stx2126.

Keane, E. F., et al. "On the Origin of a Highly Dispersed Coherent Radio Burst." *Monthly Notices of the Royal Astronomical Society: Letters* 425, no. 1 (September 2012): L71–L75. https://doi.org/10.1111/j.1745-3933.2012.01306.x.

Lorimer, D. R., et al. "A Bright Millisecond Radio Burst of Extragalactic Origin." *Science* 318, no. 5851 (November 2007): 777–780. https://doi.org/10.1126/science.1147532.

Luo, Rui, et al. "Simulating High-Time Resolution Radio-Telescope Observations." *Monthly Notices of the Royal Astronomical Society* 513, no. 4 (July 2022): 5881–5891 . https://doi.org/10.1093/mnras/stac1168.

"On the Origins of Fast Radio Bursts and Perytons: A Statement." CAASTRO. http://caastro.org/project/on-the-origins-of-fast-radio-bursts-and-perytons-a -statement.

Petroff, E., et al. "Identifying the Source of Perytons at the Parkes Radio Telescope." *Monthly Notices of the Royal Astronomical Society* (April 2015): 1–8. https://doi.org /10.48550/arXiv.1504.02165.

Simons, Daniel J., and Christopher F. Chabris. "Gorillas in Our Midst: Sustained Inattentional Blindness for Dynamic Events." *Perception* 28, no. 9 (September 1999): 1059–1074. https://doi.org/10.1068%2Fp281059.

Spitler, L. G., et al. "Fast Radio Burst Discovered in the Arecibo Pulsar Alfa Survey." *Astrophysical Journal* 790 (August 2014): 1–9. http://dx.doi.org/10.1088/0004 -637X/790/2/101.

Thornton, D., et al. "A Population of Fast Radio Bursts at Cosmological Distances." *Science* 341, no. 6141 (July 2013): 53–56. https://doi.org/10.1126/science.1236789.

CHAPTER 17. Cosmic Tremors

Aasi, J., et al. "Improved Upper Limits on the Stochastic Gravitational-Wave Background from 2009–2010 LIGO and Virgo Data." *Physical Review Letters* 113, no. 23 (2014): 1–10.

Abbott, B. P., et al. (LIGO Scientific Collaboration and Virgo Collaboration). "Observation of Gravitational Waves from a Binary Black Hole Merger." *Physical Review Letters* 116, no. 6 (11 February 2016): 061102. https://doi.org/10.1103/PhysRevLett.116.061102.

Cusack, B., et al. "Upper Limits on Gravitational Wave Bursts in LIGO's Second Science Run." *Physical Review D* 72, no. 6 (2005): 62001–62025.

"Gravitational Waves Hit the Late Show." *Late Show with Stephen Colbert*, 25 February 2016. https://www.youtube.com/watch?v=ajZojAwfEbs&t=342s.

McClellan, D., et al. "Upper Limits on the Stochastic Background of Gravitational Waves." *Physical Review Letters* 95, no. 22 (22 November 2005): 221101–221106.

Overbye, D. "Gravitational Waves Detected, Confirming Einstein's Theory." *New York Times*, 11 February 2016. https://www.nytimes.com/2016/02/12/science/ligo -gravitational-waves-black-holes-einstein.html.

Piled Higher and Deeper (PHD) Comics. "Gravitational Waves Explained." 2016. https://www.youtube.com/watch?v=4GbWfNHtHRg&t=117s.

"The Sensitivity of the Advanced LIGO Detectors at the Beginning of Gravitational Wave Astronomy." LIGO Scientific Collaboration. https://www.ligo.org/science /Publication-O1Noise/index.php.

CHAPTER 18. The Return of the Furbies

"Artificial Intelligence Helps Find New Fast Radio Bursts." SETI Institute, 10 September 2018. https://www.seti.org/press-release/artificial-intelligence-helps-find-new -fast-radio-bursts.

Chatterjee, Shami. "Focus on the Repeating Fast Radio Burst FRB 121102." *Astrophysical Journal Letters*. https://iopscience.iop.org/journal/2041-8205/page/Focus_on _FRB_121102.

Crane, Leah. "Could Fast Radio Bursts Really Be Powering Alien Space Ships?" *New Scientist*, 10 March 2017. https://www.newscientist.com/article/2124209-could-fast -radio-bursts-really-be-powering-alien-space-ships.

Pen, Ue-Li, and Liam Connor. "Local Circumnuclear Magnetar Solution to Extragalactic Fast Radio Bursts." *Astrophysical Journal* 807, no. 2 (July 2015). https://doi.org/10.1088/0004-637X/807/2/179.

Petroff, E., et al. "FRBCAT: The Fast Radio Burst Catalogue." *Publications of the Astronomical Society of Australia* 33 (2016). https://doi.org/10.1017/pasa.2016.35.

Spitler, L. G., et al. "A Repeating Fast Radio Burst." *Nature* 531 (March 2016): 202–205. https://doi.org/10.1038/nature17168.

CHAPTER 19. LIGO, We Have a Problem

Abbott, B. P., et al. "Localization and Broadband Follow-Up of the Gravitational-Wave Transit GW150914." *Astrophysical Journal Letters* 826, no. 1 (July 2016). https://doi.org/10.3847/2041-8205/826/1/L13.

Hoang, B-M., et al. "Binary Natal Kicks in the Galactic Center: X-Ray Binaries, Hypervelocity Stars, and Gravitational Waves." *Astrophysical Journal* 934, no. 1 (20 July 2022): 54. https://doi.org/10.3847/1538-4357/ac7787.

"LISA: Laser Interferometer Space Antenna." NASA. https://lisa.nasa.gov.

Martynov, D. V., et al. "Sensitivity of the Advanced LIGO Detectors at the Beginning of Gravitational Wave Astronomy." *Physical Review D* 93, no. 11 (June 2016). https://doi.org/10.1103/PhysRevD.93.112004.

Naoz, S., et al. "A Hidden Friend for the Galactic Center Black Hole, Sgr A*." *Astrophysical Journal Letters* 888, no. 1 (1 January 2020): L8–L16. https://doi.org/10.3847/2041-8213/ab5e3b.

Nieto, J-L., and Monnet, G. "Supermassive Black Holes in the Centers of Galaxies?" *AIP Conference Proceedings* 254, no. 36 (1992). https://doi.org/10.1063/1.42253.

Yoshihide, Kozai. "Secular Perturbations of Asteroids with High Inclination and Eccentricity." *Astronomical Journal* 67 (November 1962): 591. https://doi.org/10.1086/108790.

CHAPTER 20. Impeccable Timing

Detweiler, Steven. "Pulsar Timing Measurements and the Search for Gravitational Waves." *Astrophysical Journal* 234 (December 1979): 1100–1104. https://doi.org/10.1086/157593.

Jenet, Frederick A., et al. "Detecting the Stochastic Gravitational Wave Background Using Pulsar Timing." *Astrophysical Journal* 625 (May 2005): L123–L126. https://doi.org/10.1086/431220.

Lorimer, Duncan R. "Binary and Millisecond Pulsars." *Living Reviews in Relativity* 8 (November 2005). https://doi.org/10.12942/lrr-2005-7.

Lynch, Ryan S. "Pulsar Timing Arrays." *Journal of Physics: Conference Series* 610 (2015). https://doi.org/10.1088/1742-6596/610/1/012017.

Manchester, R. N. "The Parkes Pulsar Timing Array." *Chinese Journal of Astronomy and Astrophysics* 6 (2006): 139–147. https://doi.org/10.1088/1009-9271/6/S2/27.

Maselli, A., S. Marassi, and M. Branchesi. "Binary White Dwarfs and Decihertz Gravitational Wave Observations: From the Hubble Constant to Supernova Astrophysics." *Astronomy and Astrophysics* 635 (March 2020): 1–10. https://doi.org /10.1051/0004-6361/201936848.

National Radio Astronomy Observatory. "Gravitational Wave Search Provides Insights into Galaxy Evolution and Mergers." *Science Daily*, 5 April 2016. www.sciencedaily .com/releases/2016/04/160405122609.htm.

Purver, M. "The European Pulsar Timing Array." Paper presented at Planets to Dark Energy: The Modern Radio Universe, 1–5 October 2007, University of Manchester, Manchester, England. https://ui.adsabs.harvard.edu/abs/2007mru..confE.125P /abstract.

CHAPTER 21. All Together Now

Abbott, B. P., et al. "GW170817: Observation of Gravitational Waves from a Binary Neutron Star Inspiral." *Physical Review Letters* 119 (October 2017): 1–18. https://doi .org/10.1103/PhysRevLett.119.161101.

———. "Properties of the Binary Neutron Star Merger GW170817." *Physical Review* 10, no. 9 (January 2019): 1–32. https://doi.org/10.1103/PhysRevX.9.011001.

Andreoni, I., et al. "Follow Up of GW170817 and Its Electromagnetic Counterpart by Australian-Led Observing Programmes." *Publications of the Astronomical Society of Australia* 34 (December 2017): 1–21. https://doi.org/10.1017/pasa.2017.65.

"Binary Neutron Star Simulations." NCSA Gravity Group. https://gravity.ncsa.illinois .edu/binary-neutron-star-simulations.

"GW170817: Observation of Gravitational Waves from a Binary Neutron Star Inspiral." LIGO Scientific Collaboration, 16 October 2017. https://www.ligo.org/science /Publication-GW170817BNS/index.php.

Korobkin, Oleg, et al. "Gamma Rays from Kilonova: A Potential Probe of r-Process Nucleosynthesis." *Astrophysical Journal* 889, no. 2 (February 2020). https://doi.org /10.3847/1538-4357/ab64d8.

Lamb, G. P., et al. "The Optical Afterglow of GW170817 at One Year Post-Merger." *Astrophysical Journal Letters* 870, no. 2 (January 2019): L15–L19. https://doi.org/10 .3847/2041-8213/aaf96b.

"LIGO/Virgo Identification of a Possible Gravitational-Wave Counterpart." NASA. https://gcn.gsfc.nasa.gov/other/G298048.gcn3.

National Radio Astronomy Observatory. "Gravitational Wave Search Provides Insights into Galaxy Evolution and Mergers." *Science Daily*, 5 April 2016. www.sciencedaily .com/releases/2016/04/160405122609.htm.

Sanders, Robert. "Did Rapid Spin Delay 2017 Collapse of Neutron Stars into Black Hole?" *Berkeley News*, 1 March 2022. https://news.berkeley.edu/2022/03/01/did -rapid-spin-delay-2017-collapse-of-neutron-stars-into-black-hole.

CHAPTER 22. Multiple Eyewitness Accounts

"About Rubin Observatory." Rubin Observatory. https://www.lsst.org/about.

Adreoni, Igor, and Jeff Cooke. "The Deeper Wider Faster Programme: Chasing the Fastest Bursts in the Universe." *Proceedings of the International Astronomical Union Symposium* 339 (August 2018): 135–138. https://doi.org/10.1017/S1743921318002399.

"IceCube Detector." IceCube. https://icecube.wisc.edu/science/icecube.

IceCube, et al. "Multi-Messenger Observations of a Flaring Blazar Coincident with High-Energy Neutrino IceCube-170922A." *Science* 361, no. 6398 (July 2018): 1–S44. https://doi.org/10.1126/science.aat1378.

Keivani, A., et al. "A Multimessenger Picture of the Flaring Blazar TXS 0506+056: Implications for High-Energy Neutrino Emission and Cosmic Ray Acceleration." *Astrophysical Journal* 864, no. 1 (August 2018): 1–16. https://doi.org/10.48550/arXiv.1807.04537.

"Scientists Discover the Most Luminous Quasar with an Ultramassive Black Hole in the Distant Universe." Kavli Foundation, 25 February 2015. https://kavlifoundation.org/news/scientists-discover-most-luminous-quasar-ultramassive-black-hole-distant-universe.

Stein, Robert. "Tidal Disruption Events and High-Energy Neutrinos." Paper presented at the 37th International Cosmic Ray Conference, Berlin, 12–23 July 2021. https://arxiv.org/pdf/2110.01631.pdf.

CHAPTER 23. Furbies—A New Hope

Agarwal, Devansh, et al. "A Fast Radio Burst in the Direction of the Virgo Cluster." *Monthly Notices of the Royal Astronomical Society* 490, no. 1 (November 2019): 1–8. https://doi.org/10.1093/mnras/stz2574.

Burke-Spolaor, Sarah. "Multi-Wavelength and Multi-Messenger Fast Radio Burst Follow-Up." *Nature Astronomy* 2 (November 2018): 845–848. https://doi.org/10.48550/arXiv.1811.00194.

Chandra, Prakash. "Discovery of a Fast Radio Burst within the Milky Way Throws Up an Unlikely Source." *Science: The Wire*, 5 November 2020. https://science.thewire.in/the-sciences/fast-radio-burst-milky-way-discovery-chime-stare2-magnetar.

"CHIME Experiment." CHIME Collaboration. https://chime-experiment.ca/en.

CHIME/FRB Collaboration. "CHIME Outrigger Telescopes Boost Search for Fast Radio Bursts." Newsroom Institutional Communications, 30 March 2022. https://www.mcgill.ca/newsroom/channels/news/chime-outrigger-telescopes-boost-search-fast-radio-bursts-338766.

——. "The First CHIME/FRB Fast Radio Burst Catalog." *Astrophysical Journal Supplement Series* 257, no. 2 (December 2021): 1–67. https://doi.org/10.48550/arXiv.2106.04352.

Hensley, Kerry. "Exploring a Magnetospheric Origin for Fast Radio Bursts." American Astronomical Society, *Nova*, 19 January 2022. https://aasnova.org/2022/01/19

/exploring-a-magnetospheric-origin-for-fast-radio-bursts/?fbclid=IwAR3aMCUTwU
AOfPItxs0dqv4cok44yd79QNkyhH75JZA3DbDKT0hzREFSIb4.

"Milky Way Magnetar Could Be the Source of a Fast Radio Burst." *Physics World*, 4 November 2020. https://physicsworld.com/a/milky-way-magnetar-could-be-the -source-of-a-fast-radio-burst.

Newburgh, Laura B., et al. "Calibrating CHIME: A New Radio Interferometer to Probe Dark Energy." *Ground-Based and Airborne Telescopes* 9145 (July 2014): 1709–1726. https://doi.org/10.1117/12.2056962.

Petroff, E., J. W. T. Hessels, and D. R. Lorimer. "Fast Radio Bursts at the Dawn of the 2020s." *Astronomy and Astrophysics Review* 30 (March 2022): 1–49. https://doi.org /10.1007/s00159-022-00139-w.

CHAPTER 24. The First Bumps in the Universe

Abbott, T. M. C., et al. "Dark Energy Survey Year 3 Results: A 2.7% Measurement of Baryon Acoustic Oscillation Distance Scale at Redshift 0.835." *Physical Review D* 105, no. 4 (2022): 1–24. https://doi.org/10.1103/PhysRevD.105.043512.

Bailes, M., et al. "Gravitational-Wave Physics and Astronomy in the 2020s and 2030s." *Nature Reviews: Physics* 3 (April 2021): 344–366. https://doi.org/10.1038/s42254 -021-00303-8.

Bassett, Bruce, and Renée Hlozek. "Baryon Acoustic Oscillations." In *Dark Energy: Observational and Theoretical Approaches*, edited by Pilar Ruiz-Lapuente, 246–278. Cambridge: Cambridge University Press, 2010. https://doi.org/10.1017 /CBO9781139193627.010.

Beutler, Florian, et al. "The 6dF Galaxy Survey: Baryon Acoustic Oscillations and the Local Hubble Constant." *Monthly Notices of the Royal Astronomical Society* 416, no. 4 (2011): 3017–3032. https://doi.org/10.1111/j.1365-2966.2011.19250.x.

Eisenstein, D. J., et al. "Detection of the Baryon Acoustic Peak in the Large-Scale Correlation Function of SDSS Luminous Red Galaxies." *Astrophysical Journal* 633, no. 2 (November 2005): 560–574.

CHAPTER 25. The Last Hurrah

Caplan, M. E. "Black Dwarf Supernova in the Far Future." *Monthly Notices of the Royal Astronomical Society* 497, no. 4 (2020): 4357–4362. https://doi.org/10.1093/mnras /staa2262.

INDEX